A Kitchen Course in Electricity and Magnetism

David Nightingale • Christopher Spencer

A Kitchen Course in Electricity and Magnetism

 Springer

David Nightingale
New Paltz, NY, USA

Christopher Spencer
Hereford, UK

Videos to this book can be accessed at http://www.springerimages.com/videos/978-3-319-05304-2

Audio files for this book can be downloaded from http://extras.springer.com

ISBN 978-3-319-05304-2   ISBN 978-3-319-05305-9 (eBook)
DOI 10.1007/978-3-319-05305-9
Springer Cham Heidelberg New York Dordrecht London

Library of Congress Control Number: 2014939116

Printed on acid-free paper

Springer is part of Springer Science+Business Media (www.springer.com)

Preface

Many people's eyes glaze over when someone attempts to explain something even mildly technical, and if an equation is utilized—well, that's the end of the matter!

But it needn't be.

For those who would like a grounding in the basics—historical and modern—of electricity and magnetism as we experience them today, this book intentionally uses "no math" (except in Appendix F, which may certainly be omitted). It *does* allow the occasional "shorthand" definition, plus the simplest of arithmetic.

The work should be of interest to at least any of the following:

(a) the armchair reader (such as, but not necessarily, a curious retiree).

(b) an experimenter.

(c) school students, including the home schooled.

(d) he/she who asks such questions as "what is an LED light?"

(e) people taking college physics that seems too abstract.

(f) anyone whose day-to-day work involves electricity.

The emphasis throughout is on experiment and history—experimentation that can very often be repeated just with things found at home. (Hence "kitchen" in the title.) Of course, the armchair reader is not obliged to do these experiments and sometimes it's okay just to visualize them!

It's clear that we use physics constantly, and (presumably) always will. While we don't need to know any science to live comfortable lives, electricity is something worthwhile knowing the elements of—and on top of that it's fun. From bulb to LED, radio to TV, microwave oven to transistor, cell phone to photovoltaics, and much more around us—all involve electricity plus its nondetachable partner, magnetism.

So, enjoy!

David Nightingale   Christopher Spencer
New Paltz, NY   Hereford, UK

Contents

1 Home Electrostatics   1
   1.1   Static Electricity   1
   1.2   A Charge Detector   2
   1.3   Using Plastic Wrap   2
   1.4   What Has Happened   4
   1.5   Experiment: Two Plastic Strips   5
          1.5.1   What Is Happening   6
   1.6   Atoms   7
   1.7   Experiment: Bending Water   8
   1.8   Dipoles   9
   1.9   Experiment: Comb and Paper   10
          1.9.1   What Has Happened   11
   1.10   Making a Kitchen Electroscope   13
   1.11   Experiments with the Kitchen Electroscope   14
          1.11.1   What Has Happened   14
   1.12   Leyden Jar: Capacitors   16
   1.13   E Fields   18
   1.14   Experiment with Electroscope and Leyden Jar   19
          1.14.1   What Has Happened   21
   1.15   Experiment: Charging by Inducing Charges
          (May Be Omitted with No Loss of Continuity)   21
          1.15.1   What Has Happened   23
   1.16   More on Conductors and Insulators   23
          1.16.1   What Happened   24
   1.17   Lightning: Franklin's Bells and More   24

2 Current and Voltage   29
   2.1   Water Analogy   29
   2.2   Galvani's Frogs' Legs, and Volta's Experiment   30
          2.2.1   Tongue Experiment (After *Volta*)   30
   2.3   Experiment: Voltaic Cell   31

2.4   Experiment: The Voltaic Pile   32
2.5   Humphry Davy's Voltaic Pile   33
2.6   Sidebar Experiment: Electroplating   34
2.7   Experiment: Potato Battery   35
        2.7.1   What Was Happening   36
2.8   Amps, Volts, Energy, Power   38
2.9   Experiment: Current Through a Bulb   40
        2.9.1   What Is Happening   41
2.10   A Fuse   42
2.11   Making a Current Meter   43
2.12   Another Way to Get a Voltage: Seebeck Effect   44
2.13   Peltier Effect   47
2.14   Yet Another Way to Get a Voltage: Piezoelectricity   47
2.15   L.E.D.s vs. Bulbs   49
2.16   Concept of Resistance   50
2.17   Ohm's Law   51
        2.17.1   A Graph for Ohm's Law   53
        2.17.2   Experiment: Resistance of a Household Bulb   53
        2.17.3   What Was Happening   53
2.18   Equivalent Definition of Power   54
2.19   Lighting the LED   55
2.20   The Solar Cell: A (Part-Time) Battery   56
2.21   More on pV Cells (Solar Cells or Photodiodes)   58
        2.21.1   Actual Solar Cells from the Stores   58
        2.21.2   Note on Rechargeable Batteries: NiCad, NiMH, Li-Ion .   60
        2.21.3   Nickel Cadmium (NiCad)   60
        2.21.4   Nickel-Metal Hydride (NiMH)   60
        2.21.5   Lithium-Ion (Li-Ion)   60
2.22   A Charging Circuit, and a Difficulty   61
2.23   Brief History of Electrical Diodes   62
2.24   More Symbols   63
        2.24.1   Comment on the Various Uses of LEDs: 64
2.25   Series and Parallel: Water Analogy   64
2.26   Elements of Automobile Wiring   65
2.27   Current Measurements   68
2.28   Voltage Measurements   70
2.29   Resistance Measurements   71
2.30   Alternating Current and Direct Current (AC and DC)   72
2.31   Skin Effect   74
2.32   An AC Experiment with LEDs   75

3   Magnetism   77
3.1   Lodestones   77
        3.1.1   The North   80
3.2   Further View of Magnetism   80

3.3	A Kitchen Compass	81
3.4	Angle of Dip	82
3.5	Diamagnetism	83
3.6	Paramagnetism	83
3.7	Ferromagnetism	83
3.8	Shielding	84
3.9	Different Magnet Shapes	85
	3.9.1 Aurora Borealis	86
	3.9.2 Magnetic Bacteria	86
	3.9.3 Tapes and Swipe Cards	86
3.10	What Causes a Magnetic Field?	87
3.11	Oersted's Experiment	87
	3.11.1 Shape of the Field Due to a Loop	88
3.12	A Coil	88
	3.12.1 Experiment	90
3.13	Inductance (L)	91
3.14	A House Alarm	91
3.15	Experiment: Force on a Current Near a Magnet (Lorentz Force)	92
3.16	Direction of Lorentz Force	93
3.17	A Kitchen Motor	95
3.18	Adjacent Currents	100
	3.18.1 A Gedanken Experiment	101
3.19	Lorentz Force with Old TVs (Not New Ones!)	101
3.20	Hall Effect	102
3.21	Magnetohydrodynamics (or MHD)	103
3.22	Note on Microwave Ovens: A Subtle Example of the Lorentz Force	104
3.23	A Kitchen Experiment with Microwaves	106
3.24	Relative Motion of a Magnet and a Wire (Faraday's Law)	106
	3.24.1 Note on "e.m.f.s"	108
3.25	Transformers: An Example of Faraday's Law	108
3.26	Two Examples of Transformers	110
3.27	Electromagnetic Waves	113
3.28	The Electromagnetic Spectrum	114
3.29	Making a Kitchen Radio	115
3.30	Experiment: Falling Magnet	117
	3.30.1 What Was Happening	117
	3.30.2 Note on Neodymium Magnets	119
3.31	Eddy Currents, ARAGO, and a Kitchen Cooker	119
3.32	A Household Generator	121
4	**Elements of Transistors, and an Integrated Circuit**	123
4.1	First We Must Revisit the Diode!	123
	4.1.1 *n*-Type Material	125
	4.1.2 *p*-Type Material	126

4.2   The *pn* Junction   126
4.3   Experiment: Diode Graph   128
4.4   About Displays   130
4.5   Comment on LCDs   132
4.6   The Transistor   132
4.7   Experiment: Transistor as Switch   133
        4.7.1   What Has Happened   135
4.8   Experiment: Transistor as Amplifier   136
        4.8.1   Setting Up the Circuit   137
4.9   An "Absolute" Electroscope   140
        4.9.1   Using the Absolute Electroscope   142
        4.9.2   What Is Happening   142
4.10   Connection Between Fields and Potentials   142
4.11   Experiment: Charging and Discharging Capacitors   144
4.12   Integrated Circuits: The 555 Timer Chip   149
4.13   Experiment: A Metronome Circuit   152

Appendix A: Resistor Color Codes   155

Appendix B: Components   157

Appendix C: RFID—and Bar Codes   161

Appendix D: E-Ink   165

Appendix E: Touchscreens   167

Appendix F: Formulae   169

Glossary   171

Index   175

Authors Biography   179

Background

Electricity is a natural part of everyone's experience. Since the dawn of mankind, the crackling sounds of lightning discharges have frightened animals, split trees, started fires, and killed sailors on the open seas as well as golfers on their golf courses. It has also electrocuted people who have mistakenly sheltered under trees—mistakenly, because although the branches may seem to offer protection, a person under a branch, as we shall see shortly, is not safe there.

Let's look first, albeit briefly, at what earlier mankind knew about electricity.

In 600 BC, **Thales**, a Greek philosopher, had written down that *amber*, a brownish-yellow fossil resin found on some seashores, attracted pieces of straw if it was rubbed. Much later, by the 1600s, it became generally known that two pieces of amber rubbed with fur, and certain other cloths, tended to repel each other—and the same repulsion happened when two pieces of glass were rubbed with silk.

In the late 1500s **Dr. William Gilbert** (1544–1603), who was a physician to the Queen of England, noticed that many substances (like diamond, opal, sapphire, etc., as well as glass and amber)—after being rubbed—attracted tiny pieces of paper. He called substances that caused such effects *electrics*.

Curiosity flowered more strongly in the following centuries, and further experiments on these so-called electrics were being done throughout the 1700s in many countries by many different people.

Amongst them it's appropriate to mention the writer–statesman **Ben Franklin** (1736–1790) in America, the botanist–chemist **Charles Dufay** (1698–1739) in France, the ex-cloth-dyer in a London poor house **Stephen Gray** (1666–1736), the French engineer who had worked in Martinique and Brittany **Charles Augustin de Coulomb** (1736–1806), the experimenter **Michael Faraday** (1791–1867) in England—and, investigating many of the same phenomena as Faraday, the youthful would-be actor from Albany, NY, and later President of the Smithsonian **Joseph Henry** (1797–1878).

Now these experimenters knew that rubbed glass not only repelled rubbed glass, but was generally *attracted* to rubbed amber. Such observations led both Dufay and Franklin independently to the conclusion that there might be two distinct electricities, or what they called *electric charges*.

Thus, since *like* **electric charges repelled**, **and** *unlike* **electric charges attracted**, they arbitrarily called these two apparent types of charge *"positive"* and *"negative"*—words we still use.

Early experiments on repulsion and attraction were ultimately summed up by Coulomb in a basic law, in 1783—Coulomb's law[1]. He stated that the force between two charges—either attractive or repulsive—was proportional to the product of the charges and inversely proportional to the (square of the) distance between them. (For those who know about gravity, this is an "inverse square law," of the same form as Newton's—100 years earlier—law of gravitation.) If the charges were of opposite sign there was attraction, and if of the same sign then repulsion.

However, we still haven't said what electric charge is. Is it some kind of particle? Or even a fluid? And how do we quantify it?

Around the 1860s, the German all-round scientist **Hermann von Helmholtz** (1821–1894) had suggested that charge might be comprised of some unique and basic type of particle. His idea took on, and such a so-far hypothetical particle came to be referred to, especially in the German literature, by the delightful name *Helmholtzsche elementarquantum*.

We know now that this suspected fundamental particle does exist, and it has a negative charge; it is called the *electron*. (*"Elektron"* is the Greek word for amber.) Despite the fact that many scientists of the nineteenth century did not want to believe there could be any particle smaller than an atom, the electron was actually isolated by the Scottish scientist **J.J. Thomson** (1856–1940) in a series of experiments in Cambridge just before 1900.

The present-day picture is that a *negatively* charged body is said to have an excess of *electrons*, and a positively charged body has a deficiency of electrons. If bodies have neither excess nor deficiency (like everything in normal everyday life) we call them, rather obviously, *neutral*.

The charge on the electron is much smaller than the basic unit of charge that we use throughout science, which is the *coulomb*. Only if we had approximately 10 *billion billion* electrons would we say we have a coulomb. (Such a large number is often written in newspapers, for example, with commas, as in *10,000,000,000,000,000,000* but we will write it here with spaces, as in *10 000 000 000 000 000 000*.) Let's begin with a few electrostatics experiments easily done at home.

---

[1] Formulae are not used in our introductory book, but for readers who would like to see them they may be found in the Appendix on p. 169.

Home Electrostatics

1.1 Static Electricity

Everyone has rubbed a balloon on his/her shirt and watched the balloon cling to a ceiling, and maybe you have shuffled across a carpet on a dry day, only to receive a shock from the next thing touched. This particularly happens in winter, when the humidity is low.

Also, on a similarly dry day, you may have combed your hair and heard a crackling sound or touched someone and there's been a spark from your hand (Fig. 1.1).

Fig. 1.1 Shuffling across a carpet and getting zapped

D. Nightingale and C. Spencer, *A Kitchen Course in Electricity and Magnetism*,
DOI 10.1007/978-3-319-05305-9_1, © Springer International Publishing Switzerland 2015

1.2   A Charge Detector

There are various ways to detect electric charge, and in Fig. 1.2 we show a detector that can be assembled at home in a few minutes.

We take a strip of thin cardboard, or sturdy paper, about 4"–6" long, and fold it along its center line. We can then balance the cardboard at its center on a needle stuck in a cork, as shown.

If a charged balloon or comb that has been rubbed is brought near to either end of the charge detector, the cardboard will swing around and follow, clearly attracted, and never repelled. However, it is <u>only</u> a detector, and it will not tell us anything about the charge that is causing it to swing.

In the following pages we look at some different kitchen phenomena involving electrostatic charges and, after studying the atom, begin to explain them.

Fig. 1.2 A piece of thin cardboard is folded along its center and supported by a needle stuck in a cork. It will swing round towards any electric charge

1.3   Using Plastic Wrap

In Fig. 1.3 below, a piece of <u>un</u>charged plastic wrap hangs freely. The home experimenter will find it quite a challenge to have this uncharged, because normally, when plastic wrap is torn from its box, friction will give it a charge. In the

Fig. 1.3 A piece
of uncharged plastic
wrap hangs freely

Fig. 1.4 (**a**) If the plastic
has been torn recently,
or dragged across a dry shirt
or other cloth, it will cling
to almost anything, such as
the metal refrigerator or
(**b**) perhaps a stool

a

b

photograph, the plastic was actually unrolled slowly and carefully, and cut with sharp scissors, avoiding friction or scraping.

However, as we said, if it has been rubbed or torn off recently, it <u>will</u> be charged and will cling to almost anything, as shown in Fig. 1.4a, b.

1.4   What Has Happened

When the plastic was torn too vigorously from its box, electrons were knocked either off the surface of the plastic (leaving the plastic somewhat positive) or off the box and onto the plastic (thus leaving the plastic with an excess of electrons).

As yet we haven't given any reason which of these it is. It would actually be possible later to find out, but for simplicity let us assume now that the plastic wrap is (−), *i.e.*, too many electrons.

The plastic can cling to anything, because then (+) charges in various nearby objects will try to get close to the plastic's (−)—from the results of those early experimenters who saw that *unlike charges always attract each other and like charges repel.*

Of course, the metal and/or the nonmetal stool were originally electrically neutral. So what was going on?

Consider the refrigerator first. In the left sketch of Fig. 1.5 the plastic is neutral, as is the refrigerator. (For simplicity we have not bothered to show the equal number of pluses and minuses on the plastic, although they are shown on the refrigerator.) In the right-hand figure the (−) charges of the refrigerator have now moved as far away as possible from the (−) charged plastic wrap, leaving (+) charges nearer to the plastic wrap. Of course, there is then mutual attraction.

The refrigerator is made of metal, which is a *conductor*, and *electrons are free to move in a conductor*. We will take this as the definition of a conductor (and discuss semiconductors later).

In the case of the kitchen stool (see Fig. 1.4b), which is not a conductor but rather an *insulator* (we use the word insulator for a material whose charges are NOT free to migrate from one place to another) something has obviously happened that there is still attraction.

Note that in these kitchen experiments the plastic wrap may eventually lose its attraction because some of the charges on the surface of the plastic may very slowly get neutralized, perhaps by actual contact with the metal refrigerator, and/or from

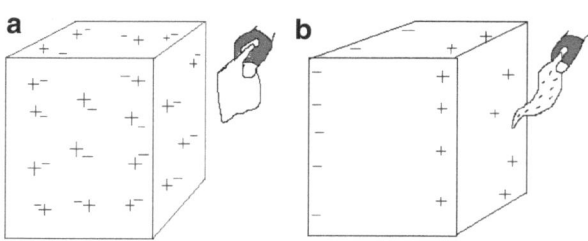

Fig. 1.5 (a) Original uncharged plastic, (b) charged plastic

Fig. 1.6 Plastic wrap still clinging weakly. (This has been here for over 3 months . . .!)

stray atoms in the air that may have lost or gained an electron. (Such atoms and/or molecules are called *ions*.)

Our Fig. 1.6 shows a piece of charged plastic still clinging to the side of a metal object after many days!

1.5   Experiment: Two Plastic Strips

Cut two strips of plastic, as shown in Fig. 1.7. It's best if these have a certain rigidity, and we have colored them only so as to render them visible against the white background.

Rub them all the way down with your bare fingers or with a sock or some kind of wool—just once or twice; this will make the strips, which are identical, repel each other, as on the left side of Fig. 1.7.

Next, bring up a kitchen knife, or some metallic object, between the strips and witness them both collapse onto the metal, as on the right of Fig. 1.7.

Take the knife away, and they will go back to repelling each other again.

Fig. 1.7 The charged strips repel each other, but if a conductor is brought between them they will collapse onto it. When the conductor is removed, the strips repel each other once more

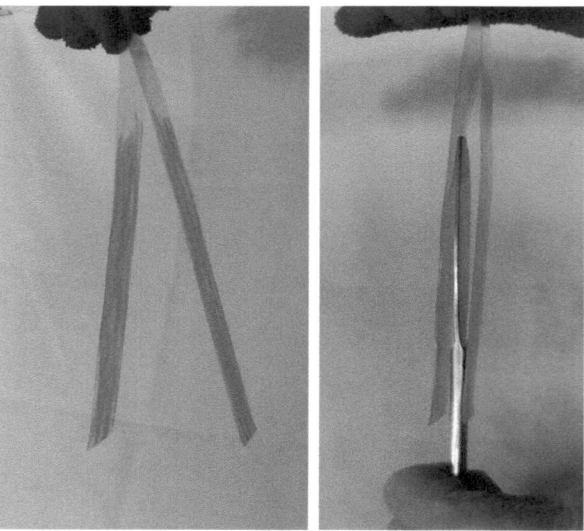

1.5.1 What Is Happening

The plastic is an insulator (defined on p. 4). By friction some charges were knocked off.

There must be like charges on the two pieces of plastic for them to repel.

When a conductor is brought between them the opposite sign of charge is attracted as close as possible to the charges on the plastic strips. Thus each strip is attracted to the metal—and clings to it.

When the metal is taken away, the strips return almost to their mutual repulsion. We say "almost" because there may have been a small amount of cancellation from the direct contact.

In order to understand better the previous experiments we must now digress and describe the basics of an atom.

However it should be emphasized that we have never clearly seen an atom, although photographs of "lumps" representing arrays of atoms exist now, and may be seen in modern physics texts (*e.g.*, E. Hecht's "*Physics*," on his p. 318.)

The Greek philosopher **Democritus** (roughly 460–370 BC)—known as the "laughing philosopher" on account of his emphasis on cheerfulness—was one of the first to believe that matter consisted of atoms. His idea was that there were just two things—atoms and void. His atoms were hard and indivisible and of different shapes and sizes, and he suggested that they could form clusters of distinct types—a precursor, perhaps of today's *molecules*.

One of Einstein's earliest papers ("*On Brownian Motion*," 1905), explaining the mathematics of pollen (on the surface of water) being hit from all sides by unseen particles, helped support the belief in Democritus' idea that matter had to be made up of atoms.

1.6 Atoms

The model of the atom commonly accepted today is due to the Danish scientist, **Niels Bohr** (1885–1962). Because we are in the kitchen we will not need to discuss the more complicated quantum mechanical model of the atom, and we will be at no disadvantage here by such omission, because Bohr gave us the essential picture.

Atoms have a central massive (+) nucleus, orbited by electrons—like planets around the sun. However, instead of gravity, the electrons are attracted to the nucleus as dictated by Coulomb's law (p. xii).

The positive particles in the nucleus are called *protons*, of equal and opposite charge to the tiny electrons—but much heavier (Fig. 1.8).

The number of protons indicates the identity of the atom. For example, the hydrogen nucleus has 1 proton, helium has 2 protons, and uranium has 92 protons.

A *special rule* controls the atom's chemical behavior. Let us look at an atom of carbon. Carbon has a total of six protons, so it must have six electrons. Two of these electrons are in its first "shell" or "radius" or "orbit," and according to the special rule—which comes from quantum mechanics—it is not allowed any more in that shell.

So the remaining four electrons have to be found in the second shell. (If we continue the rough comparison to astronomy, in our solar system Mercury would be in the first shell, Venus in the second shell, Earth in the third, and so on.)

Now carbon has a strong chemical resemblance to other atoms that also happen to have only four electrons in their outermost shell—a prime example being silicon, which has four electrons in its <u>third</u> shell. We will meet silicon in Chap. 4 when we deal with transistors.

As we said, the proton is much heavier than the electron—1836 times heavier.

In the nucleus there are also neutral particles, called *neutrons*. These will not concern us in our study of electricity, and we may regard them just as "ballast." The nucleus is thus even heavier than might be expected if all it had were protons.

The whole atom itself is "neutral," *i.e.*, having no net charge, so again this means that if there are 17 protons in the nucleus there must be 17 orbiting electrons in their various shells.

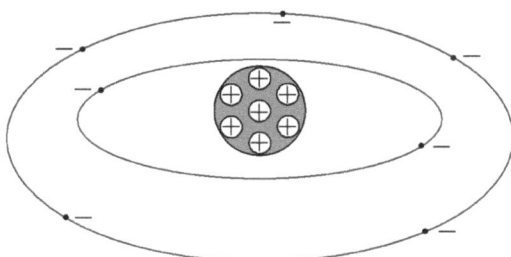

Fig. 1.8 The Bohr atom. Like planets around the sun, tiny electrons orbit the nucleus. There are also neutrons in the nucleus, which we haven't shown. This particular atom has seven electrons and is therefore nitrogen (see the Periodic Table in any encyclopedia)

We should mention that for any element, different neutron numbers are possible, and these atoms are called *isotopes* of the element (from the Greek *iso*, same, followed by "*p*" for protons—same number of protons.) For example, looking again at carbon, the most common form has six protons as well as six neutrons in its nucleus. It is customary to write it as $^{12}_{6}C$. A well-known isotope of carbon, $^{14}_{6}C$, has two extra neutrons. (This isotope is also *radioactive*, which means that its nucleus is slowly disintegrating, or rather changing, by radiating off particles.) Carbon 14 is extremely useful in dating ancient organic artifacts in archeological excavations.

The special rule that we referred to which describes the maximum number of electrons allowed in each radius or "shell" is called the *Pauli exclusion principle*, which, again, we don't discuss in the kitchen! The only point we need to remember is that there are these limits. Shells are regarded as "filled" when by that rule they are not allowed to have any more electrons. We may just mention that the number of electrons allowed in the second shell is eight, but after that the numbers become more complicated—as chemists know—and we will not pursue this here.

Finally, we mentioned that if we were to take an orbiting electron away, the atom would become *positive*, and scientists would then call it a *positive ion* rather than an atom. Similarly, if we were to somehow add an electron, the atom would obviously become negative and would be referred to as a *negative ion*.

In atoms that have many electrons some of the outer orbiting electrons are relatively easy to knock off by friction, which is what was happening in the "rubbing" examples with the plastic sheets above.

All those early experimenters who lived prior to 1900 would have been interested to know that it's just the transferring of electrons that is responsible for the effects they were observing.

The addition or the subtraction of electrons today is central to our understanding of many appliances in modern life. Electrostatic effects can be used, for example, in laser printers and photocopiers, where tiny sections of *positively* charged paper can attract sootlike "toner" particles—on the assumption that the particles have been *negatively* charged. (The resulting "image" can be "fixed," or made permanent, by applying heat.)[1]

Earlier, we noticed that our plastic wrap clung to an *insulator*, such as the stool. We will also see, in the following experiment, that something like regular water seems to be affected by electric charge. Now that we know the basics of an atom we will be able to explain these things.

1.7   Experiment: Bending Water

Turn on the kitchen faucet, and adjust the flow to a very thin continuous trickle. Run a dry comb through your dry hair (or rub the comb with wool), and bring the comb close to, but not touching, the stream (Fig. 1.9).

---

[1] One can find many further industrial applications described in technical encyclopedias.

Fig. 1.9 Bending water

The water has been attracted to the charged comb, and we'll see the reason after first describing a *dipole*.

1.8   Dipoles

By the definition we gave on p. 4 charges do not travel in insulators, nor in things like pure water. However, if the molecules are *dipolar*, they can act like the needle of a compass, and <u>swing</u> around, without leaving their position. We illustrate this below.

It is possible for one side of a molecule to be more positive (or negative) than the other, depending on the arrangement of the atoms in the particular molecule. Instead of N and S, as with a common magnet (to be studied later) some molecules may have (+) and (−) ends, as shown below in Fig. 1.10.

It is common for everyone to refer to water as H_2O, which signifies that there are two hydrogen atoms attached to one oxygen atom to make a molecule of water.

In Fig. 1.10 the electrons of the oxygen atom have an "average place" somewhat to the top left of the diagram.

Also, the region of the two hydrogen atoms is slightly more positive than if they were alone because their electrons (just one each) are actually being shared by the oxygen. Why? Because, by that rule we mentioned concerning "filled shells," the outer shell of the oxygen would prefer to have a complete set of eight electrons, and the two hydrogen atoms, with one electron each, provide just that.

The net result is that the water molecule is electrically rather like a dog bone, (+) at one end and (−) at the other, or analogous to a magnet, except that we're not talking about magnetism here. We call the water molecule *dipolar*.

<u>Note</u>: Substances that may not be inherently dipolar can also act in the same way as the water molecule. Figure 1.11 shows a charged comb <u>distorting</u> the electron cloud in a nearby atom, making the atom again rather like a "dog bone" or a dipole.

Fig. 1.10 A water molecule as a "dog bone" (or *dipole*). The "H"s constitute the (+) end, and the O is more negative, as explained below

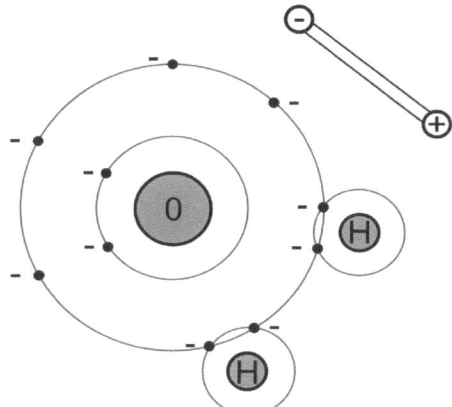

Fig. 1.11 A charged comb repelling an atom's electrons so that it acts <u>like</u> a dipole

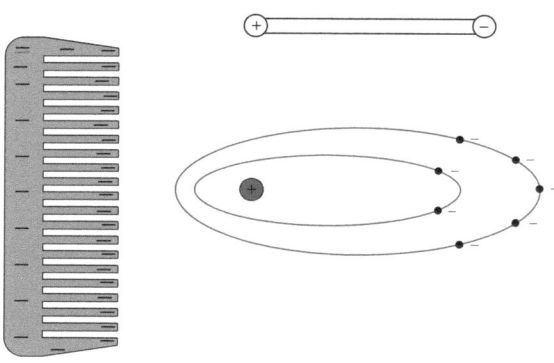

This dipolar nature explains why the plastic wrap clung to the stool as well as why the water bent. The molecules in the stool and the water swung around and lined up, such that the (+) ends of the dipoles were attracted to the (−) of the comb or the plastic wrap.

Two more illustrations follow.

1.9   Experiment: Comb and Paper

Tear some tiny pieces (we mean tiny!) off the corner of the newspaper. Alternatively, use some pieces of Styrofoam, perhaps from packaging or a cup. In either case, they must be lightweight.

Then take the ordinary plastic comb, rub it again through your dry hair (or with wool), and note how it will pick up the smidgins of paper or Styrofoam. This of course is also exactly what happened with our paper charge detector on p. 2.

Further, if you wait for as much as a minute, some of the little pieces may suddenly fly off!

(Note again: ELECTROSTATICS EXPERIMENTS WORK MUCH BETTER IN THE WINTER WHEN THE HUMIDITY IS LOW.)

1.9.1   What Has Happened

If the comb were negative (say) the (+) and (−) charges of the molecules of the neutral paper became distorted, acting as dipoles. The (+) ends were attracted to the (−) comb, as shown in Fig. 1.12. And, as we said, exactly the same thing was happening with our paper "charge detector" on p. 2, where we put off the explanation.

And why did the tiny pieces of paper decide to jump off the comb a few moments later?

By contact. As they rested on the comb some of the smidgins gradually acquired, by direct touch, a negative charge, just like the comb; hence—poof!—repulsion.

Table: While doing experiments that involve charges obtained by rubbing, we should give here a list of substances which, if below in the list, are more negatively charged than the substances above. For example, frictional contact between *glass* and *silk* will leave the glass more positive.

Rabbit fur

Glass, mica, wool

Cat fur, lead, silk

Aluminum

Cotton, wood, amber, brass

Rubber

Celluloid

India rubber

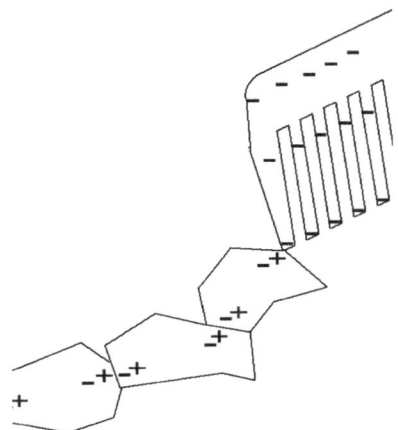

Fig. 1.12 Arrangement of charges for (−) comb picking up paper or Styrofoam

Fig. 1.13 Plastic spoon
and hanging plastic wrap.
(For clarity, the background
is *white* and is not touching
either plastic or spoon)

As an exercise, what do you think is happening in Fig. 1.13.

Answer: The plastic wrap is being repelled by the comb/spoon, and so they must
have the same charge "polarity," either both (+) or both (−).

The *quantity* of charge (number of Coulombs) on each need not be the same.

As another exercise, this time about dipoles, might we not be tempted to think
that the cell or the battery in Fig. 1.14 is a dipole? If so, shouldn't it swing round?

Answer: The cell certainly has its (+) and (−) at opposite ends, as labeled.

However, these cells are constructed in such a way that underneath the label is *a
metal case along the whole cylinder*. (Scraping off the label showed that an old
alkaline cell had its metal case (+), whereas a Fuji rechargeable had its case (−).)

Thus, tempting as it may seem, ordinary household batteries do not behave like
dipoles.

Furthermore, 1.5 V is a very small voltage compared to the voltage that could
be found on a charged comb or spoon—IF this could be measured by an instrument that draws no current. The home experimenter does not have access to
such an instrument (usually called an *electrometer*—see glossary on p. 171).
However, the instrument we are about to study, the *electroscope* would be the
closest to such a device.

Fig. 1.14 Comb and battery

1.10 Making a Kitchen Electroscope

Suppose we want to know roughly <u>how much</u> a comb is charged.

An *electroscope* is an instrument that not only shows this but also gives a rough indication of how large the charge is. It is easy to make such a device with what can be found at home, and here (Fig. 1.15) is one that was working within 15 min.

<u>Needed</u>: *A clean dry glass jar (we want to see through it), any old plastic lid (it can overlap the jar), a paper clip, and two small strips of lightweight aluminum foil.*

Fig. 1.15 Kitchen electroscope: Aluminum foil hanging from paper clip

It is only necessary to prick a pin hole in the plastic lid, reshape the paper clip (as shown), and push it through the hole (bending it again). The hole should be tight enough to stop the clip from sliding down loosely.

Let two slivers of aluminum foil hang inside the jar from the paper clip. (Of course you will have to prick a hole in the end of each sliver.)

By the way, such home electroscopes can be just as useful as those costing $100–$300 from science lab suppliers.

1.11 Experiments with the Kitchen Electroscope

At the risk of too much repetition, we must emphasize that electrostatics experiments work much better when the indoor humidity is below ~35 %. They are absolutely not worth attempting on humid summer days!

Make sure that the two "slivers" or leaves of aluminum foil are hanging freely.

Also, gently touching the paper clip of the electroscope with a moistened finger will reset (ground) the electroscope, in case there is any stray charge causing the leaves to be slightly apart. Electrons will flow to (or from) the essentially infinite ground and neutralize things before we start.

Then:
1. Run a comb through your dry hair (or rub it again on your sleeve), and bring the comb slowly towards the top of the paper clip. Do not actually touch the paper clip. You should see the leaves diverge as the comb approaches and die down again as you take the comb away.
2. Reset the electroscope to zero (if necessary) with your moistened finger. Rub the comb again, and now balance it carefully on the top of the electroscope (*i.e.*, the bent paper clip) as in Fig. 1.16. The leaves will repel each other, probably more strongly than in (1). They should stay that way for quite a time, depending on the day's humidity.

1.11.1 What Has Happened

Let's assume that the comb has (say) a negative charge.
1. When the comb approaches the initially neutral electroscope, the (−)s in the metal (paper clip-plus-foil assembly) are repelled as far away as possible, leaving the (+)s close to the comb. The charges are simply obeying the basic law enunciated by Coulomb—that unlike charges attract and like charges repel. So we are left with predominantly (−) charges on the two "leaves," and the leaves must diverge, as sketched in Fig. 1.17.
2. When the (−) comb actually touches the top of the electroscope, the relative (+) at the top of the paper clip is partly neutralized. (Note that the (+)s don't actually migrate in solids—it is the electrons that move.) The electrons are again repelled to the leaves, but because some of the (+)s were neutralized a net (−) charge now resides on the electroscope. When the comb is taken away, the electroscope still has that extra (−) charge—it can't go anywhere—and so the leaves will remain apart.

Fig. 1.16 Charged comb on electroscope

Fig. 1.17 Leaves diverging

If comb is (say) (−) . . . .

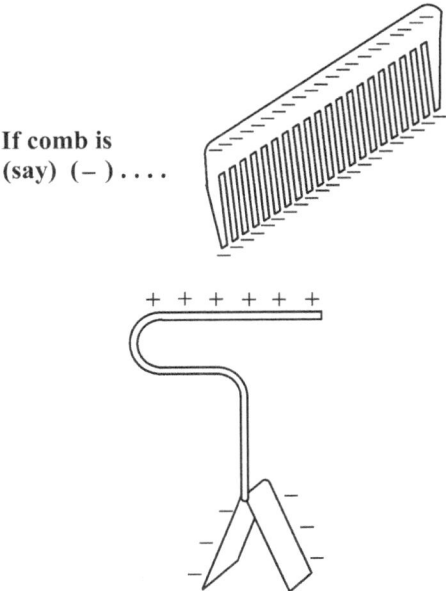

If you repeat the above experiment a few times, you may be able to build up the divergence of the leaves, each time neutralizing more (+)s.

In principle, our electroscope could be calibrated to give the amount of charge, *i.e.*, the actual number of electrons gained (or lost)—measured in Coulombs— because a large divergence means many coulombs of charge. Thus the electroscope is a rough-and-ready "charge meter."

Charge, by the way, is directly proportional to what the Electricity Company "charges" you for (same word, similar significance)—but they don't use an electroscope to measure what you've used! They commonly use an electromagnetically rotating wheel that one can usually see behind the glass of the utility meter.

A charged comb, or any charged object, possesses energy. We come back to energies later, but how much energy we have is simply proportional to how much charge there is. In fact, we will see that energy is measured in Coulombs at a certain pressure—"Coulombs multiplied by Volts."

This Coulomb–Volt is called a "*Joule*," in honor of **James Prescott Joule** (1818–1889), son and grandson of English brewers and a brewer himself, who found the relationship between mechanical work and heat by measuring the tiny increase in temperature of water being stirred by a paddle wheel. He was a lifelong experimenter, and although his results were initially met with skepticism by other scientists he also measured how much heat was dissipated when an electric current passed through a resistor. Nevertheless, some years before he died he was made a Fellow of the Royal Society and awarded a government pension.

We will address *Volts* and *Joules* and other definitions in Chap. 2.

1.12 Leyden Jar: Capacitors

How does one keep and store electrostatic charge? Figure 1.18 shows one of the earliest methods—a glass jar used by experimenters in the Dutch city of Leyden in the 1750s. It is simply a jar with foil on the inside as well as on the outside. Opposite charges will "glare" at each other through the separating wall of glass and, because they are strongly attracted, will remain there.

A Leyden jar is sometimes called a *capacitor* or, in older parlance, a *condenser*. It has the "capacity" of being able to store a certain quantity of charge. The symbol for capacity is C. A large Leyden jar can store equal and opposite charges for quite long periods.

A Dutch professor, **Dr Pieter van Musschenbroek** (1692–1761) at the University of Leyden, the Netherlands, is reputed to have been the inventor of the jar. Soon after, an abbot, by the name of **l'Abbe Nollet**, apparently demonstrated this stored electricity to King Louis XV in Versailles, by having 180 of the king's soldiers standing in a chain holding hands. A soldier at the end held one terminal of a wellcharged Leyden jar, and when the last soldier touched the other terminal, all the soldiers jumped involuntarily. The king asked for this experiment to be repeated in Paris, and it is reported that 700 of the Abbe's monks apparently all leaped simultaneously into the air.

Fig. 1.18 A Leyden jar, for storage of electric charge. One conductor would be (+), and the other (it doesn't matter which) would be (−)

Glass separates the 2 conductors

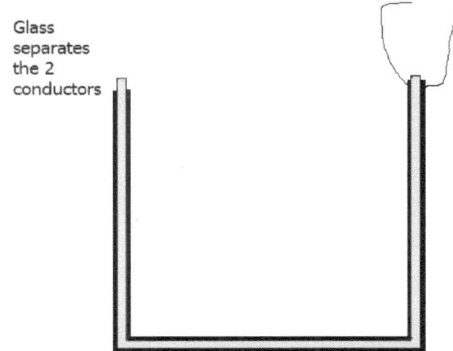

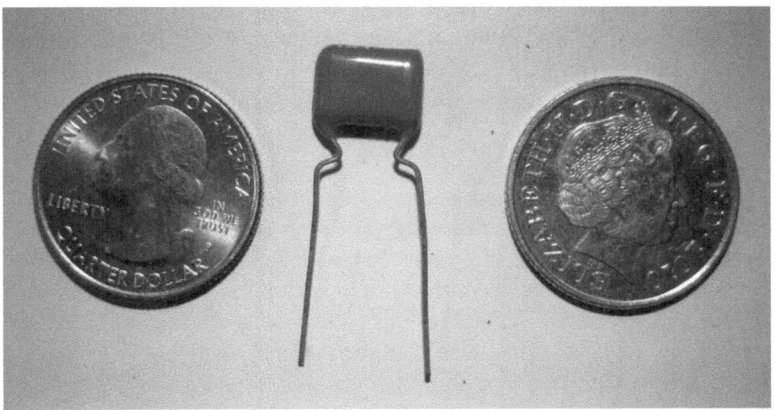

Fig. 1.19 A typical small capacitor between a US 25 cents on the *left* and a UK 10 pence, both similar in size to the Euro 50 cents (not shown)

Capacitors are used extensively in all radios and TVs and amplifiers, and a typical one is shown in Fig. 1.19. In many devices they can be many, many times smaller than this blue one. The value of a capacitor is measured in *Farads* (after Michael Faraday; see p. 106).

Capacitors have a further important use, separate from storage. Although no charge can travel from one side of a capacitor to the other (unless there's a breakdown or a spark) capacitors can allow <u>changes</u> of charge to be "transmitted," because a charge on one plate immediately induces an opposite charge on the other plate. So if a signal is to be sent along a wire without any actual charge moving between (*i.e.*, across) the plates, the capacitor will allow this. We will visit this again on the section concerning AC/DC, p. 72.

1.13   E Fields

If a tiny (+) charge (called a *test* charge) were placed near to a big fixed (+) charge, it would experience a force. It will be repelled, and Coulomb's law tells us that this force depends upon how big it is as well as how distant the tiny charge is from the fixed charge.

Exercise: Sketch some possible paths the test charge would take, just by following the above statement.

Answer: The tiny (+) test charge will be repelled from the large fixed (+) and will go off towards infinity. Many possible paths (sometimes called *"lines of force"*) are sketched below in Fig. 1.20. Such an array of possible directions is generally referred to as an "**Electric field**," or **E** field, *and it emanates from the central (+) charge in all three directions.*
    The actual force experienced by the little test charge will be strongest where the lines of force are most concentrated.

Note: There is a famous law in elementary electrostatics called *Gauss' law.* We do not need it in our introduction, but for completeness we mention it here. Visualize an imaginary balloon that has no effect on anything, enclosing the big central (+) charge.
    The electric field lines penetrate through this imaginary balloon, and it was shown by the German mathematician, astronomer, and theoretical physicist, **Johann Carl Friedrich Gauss** (1777–1855), who had shown great precociousness in mathematics as a child, that the outflow of those lines was directly proportional to the big charge in the middle. This is his almost 200-year-old law, and it is useful for those who study fields mathematically. In the kitchen we do not!
    Let us return to a Leyden jar and how it has the "capacity" to store charge.

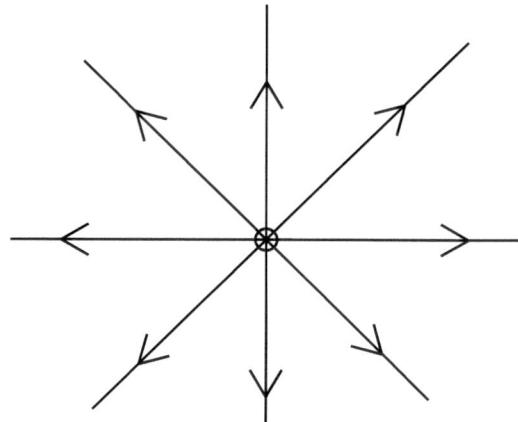

Fig. 1.20 Possible paths for a tiny (+) test charge near a fixed (+) charge. (The *lines* come out in all directions)

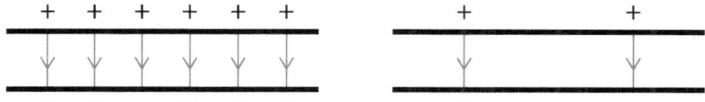

Fig. 1.21 On the *left* there is a charge of 6 coulombs. On the right we have reduced the voltage across the capacitor, and there is now only a charge of 2 coulombs

Consider Fig. 1.21. The usual symbol for a capacitor is two parallel lines, as shown by the two horizontal black lines. We have also shown the direction of the **E** field, if the top plate is positive—as already described briefly for Fig. 1.20.

Changing the charge on one plate will induce an equal and opposite charge on the other plate, because the capacitor as a whole remains neutral. The charge might be changed by the voltage we choose to apply across the plates. In radios and TVs this kind of rapid change goes on continuously.

In Fig. 1.21 the electric field, which by convention starts as we said before on the (+) and ends on the (−), here points downwards. (One cannot "see" an electric field; it is merely a useful way of knowing the direction in which a free positive "test" charge would move, <u>if</u> there were one between the plates (there is not).)

Finally, the reader might wonder if it is possible to have a charge on one plate and no charge on the other plate of a capacitor; after all, didn't we place charge on the electroscope—where there was seemingly no other plate? The glib answer is that the "other plate" was the ground, but this needs some discussion.

An initially uncharged capacitor, not connected to anything at all, might have a charged comb touch one plate. Then, by contact, a few electrons (say) could stick on that plate. If the other plate is <u>truly isolated,</u> all that would happen is there would be a local rearrangement of the charges on that neutral isolated plate, but no net charge can reach there from the outside.

For all practical purposes, however, with a capacitor in an actual circuit, there cannot be a totally isolated neutral plate, and so adding 117 charges to one side will cause 117 opposite charges to appear on the other plate via the external circuitry and wires to which it is connected.

Note that all conducting objects act as capacitors, because inevitably there will be another conducting object nearby, even if only the ground beneath.

1.14   Experiment with Electroscope and Leyden Jar

First, construct a Leyden jar by wrapping kitchen foil once around the outside of a parallel-sided glass, such as a beer mug, and secure the foil in place with a piece of ordinary transparent household tape. The inside plate is also foil, but is slightly fiddlier to make.

Our completed Leyden jar is shown on the left of Fig. 1.22.

We will need to be able to lift out the inner plate of our Leyden jar by hand—a rather delicate operation, if we are to avoid crumpling the foil—and so we have

Fig. 1.22 The inner plate on the *left* is connected to the top of the electroscope by a thin strand of copper wire, just visible at the top of the sheet that was used for background

taped a piece of dental floss (a good insulator) for a "handle." Make sure that you are satisfied that the inner cylinder of foil will slide up and down without catching on anything.

After checking that the kitchen electroscope is working properly, connect the inner plate of the Leyden jar to the top of the electroscope, using a fine strand of copper wire—barely visible in our photograph (Fig. 1.22). Ordinary connecting wire is too cumbersome for when we wish to slide out the inner cylinder. (The authors managed to slide a strand from ordinary untwisted stranded wire—not before having a few strands break off however!)

Charge the inner plate of the Leyden jar using a comb as before, touching the top of the electroscope or the inner foil of the Leyden jar (they are connected by the fine wire) at least half a dozen times. Remember that the charge is being shared with the electroscope, so we will need more charge. The electroscope slivers—or leaves—will gradually begin to show a small divergence.

This small divergence is enough, because when we slide the inner cylinder up and out of the jar the leaves will diverge, as we see in Fig. 1.23. When we slide it back down, the leaves return to their much lesser divergence. We can do this many times, losing nothing—if it's a dry day!

Fig. 1.23 Here the inner plate has been removed from the Leyden jar, and so essentially all the shared charge is now on the electroscope. (The strand connecting the two is just visible at the *bottom left* of the window)

1.14.1 What Has Happened

The charge on the inner plate of the Leyden jar was held there by the attraction to the (opposite) charges of the Leyden jar's outside plate, which was also connected to the electroscope's leaves.

When the inner plate was slid up and out of the jar, those charges on the inner plate had nothing to hold them there, and so they moved away and spread as far as possible along the strand of wire to the electroscope.

This caused the leaves to get even further apart. When the inner cylinder was reinserted, the leaves correspondingly collapsed again to the original small separation shown in Fig. 1.22, because they are once again shared, as at the start.

1.15   Experiment: Charging by Inducing Charges (May Be Omitted with No Loss of Continuity)

This experiment shows how we may *induce* a charge, not actually touching any charged object such as a comb to the electroscope.

Fig. 1.24 Charging by induction. (**a**) If (−) comb, then (+) at top, (−) on leaves, (**b**) finger approaching top (+)s, (**c**) (−)s can now escape to ground; leaves collapse. (**d**) Now only (+) charge remains

Position the charged comb *very* near to, but not touching, the top of the electroscope (Fig. 1.24a), and then gently place a moistened finger onto the "far" side of the paper clip, as in Fig. 1.24c.

Take the finger away, and then the comb, as in Fig. 1.24d.

If the experiment has been performed properly the leaves of the electroscope will remain separated, as in Fig. 1.24d.

1.15.1 What Has Happened

What are the electrons doing?

In metals, electrons are the only particles free to move. The (−) comb repelled them as far away as they could go. So the top is (+), and the leaves are both (−).

When the finger made contact it provided a path for the electrons to move even further away from the (−) comb, so some escaped through the finger to the earth, leaving the leaves to almost remain neutral. (The reader may object that electrons on the leaves would have to "climb back" towards the (+)s to get to the "finger escape route," and a few may indeed not be able to do this.)

When the finger is taken away there is no longer a connecting path to ground, and so the electroscope has an overall shortage of electrons. Atoms short of electrons now find themselves distributed more or less uniformly throughout the metal parts of the electroscope, and so the leaves diverge, indicating their (now) relatively (+) charges.

1.16 More on Conductors and Insulators

Try charging the electroscope via a reel of insulated wire connected to the top of the electroscope, as in Fig. 1.25. (A reel of loudspeaker wire would do, or other.)

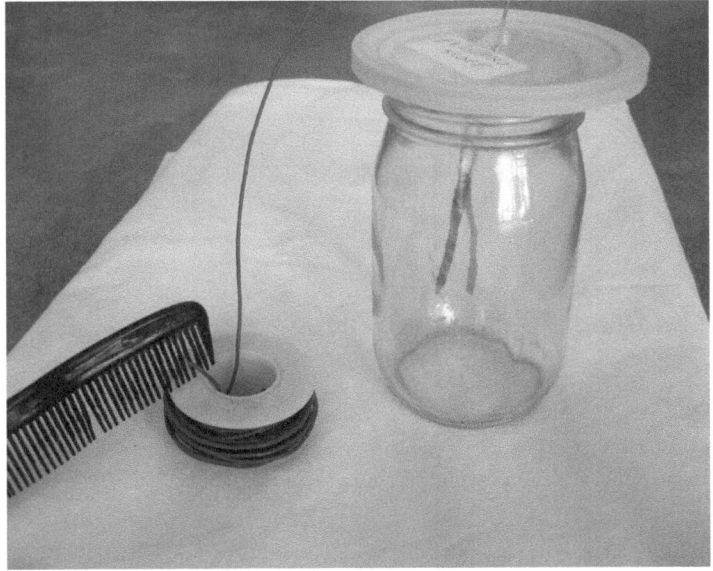

Fig. 1.25 The charge is conducted to the electroscope along the whole reel of wire

1.16.1 What Happened

This may seem obvious. Electrons on the comb easily repel the wire's electrons to the other end of the wire, because, as mentioned, electrons are free to move in most metals—finding it easiest in silver, copper, and gold, but less easy in others, such as iron. (Referring to p. 124 we see that the latter is a poor conductor, by a factor of ~7 compared to silver—so utility companies would not use iron or steel for power lines.)

One may wonder what would happen if we replaced the reel of wire in Fig. 1.25 with an insulator, like a wooden ruler, or a length of (dry) string. The result is a little surprising, and this is exactly what **Stephen Gray** living in a London poorhouse tried in 1729. He found that even with 50 feet of dry string (replacing the wire of Fig. 1.25) an electrical pulse was just transmitted!

What was happening was that although there was no conduction of any charge there was a polarization, all the way along the insulator. The home experimenter should try this, attaching a short length of string to the electroscope (letting it hang, so there is no "grounding") and touching a vigorously rubbed comb to the other end of the string. Be warned, however: the electroscope leaves will move only slightly, because the effect is very weak.

Nature's elements and compounds have much variety. In general, the outer electrons of metals are not tightly bound, and they are the reason for conduction. Moving charges are called *currents*, and we look at some kitchen experiments with currents in Chap. 2.

Finally, in reality, electrons do not actually move very long distances in a wire, but instead take short zig–zag paths, colliding here and there. The reason a device turns on apparently instantaneously (actually closer to the speed of light) from a switch far away is somewhat like a tap immediately pushing water out of the end of a filled pipe. The water molecules near the tap do not suddenly zip along to the other end of the pipe; they merely push on the neighboring molecules, forcing the end ones out with negligible delay.

The subject of *semi*conductors will be addressed in Chap. 4. We close this introduction to static charges with some comments on lightning, which involves enormous charges suddenly discharging—often, of course, doing severe damage.

1.17 Lightning: Franklin's Bells and More

To give a few random examples: in 1902 lightning destroyed an upper section of the Eiffel Tower; in 1971 lightning struck a wing fuel tank on LANSA Flight 508 in Peru, killing 91 people[2]; in 1994 lightning struck some ground fuel tanks in Egypt, killing 469 people; and 30 cows died sheltering under a tree in 2004 in Denmark.

---

[2] Astonishingly, there was one survivor, a 17-year-old German high school student who fell two miles, still strapped to her seat, into the jungle. Despite a broken collar bone and an eye injury, she managed to follow a stream, ultimately finding a loggers' camp. "Wings of Hope" (2000) is a TV documentary about the ordeal.

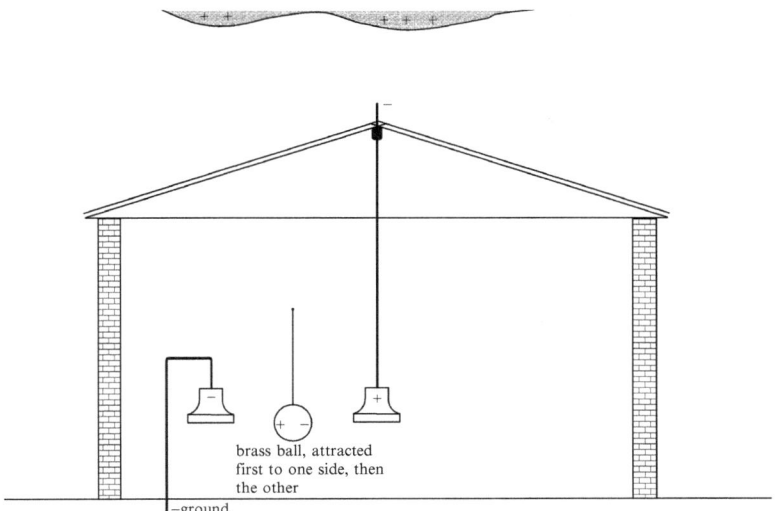

Fig. 1.26 Ben Franklin's thunder cloud warning bells, Philadelphia, 1752

There are many different classifications of lightning—ball lightning, sheet lightning, and so on—but the commonality is the separation of (+) and (−) charges, which, if they get too large, may "discharge" suddenly. (The same happens, but in a controlled way, with an automobile spark plug.) Since gases and liquids are FLUIDS what we mean by "discharge" is that, if there is a high concentration of oppositely charged regions, individual charged atoms and molecules are free to move about in what is quite often an explosive release. Another example of a discharge would be in the gas of a camera flash tube. In SOLIDS there is a fixed framework of atoms constraining any possible discharges.

In Figs. 1.26 and 1.27 we have assumed that the charges in a cloud have become separated, possibly due to violent updrafts in hot weather or atmospheric instabilities.

Let's say that this "polarization" of charges in a storm cloud is vertical—which makes sense for updrafts—with the lower part of the cloud positive and the upper part negative.

Benjamin Franklin, following a 1742 German device that he had read about, set up in his house in Philadelphia a pointed conductor sticking above his chimney, with the lower end attached to a bell inside the house. He wrote: "... *in 1752 I erected an iron rod ... about nine feet above the chimney ... down into my house ... and ... the bells rang when there was a [thunder] cloud over the house.*"

He reasoned that the top of his conductor would be of one polarity (see Fig. 1.26) and the bottom end of the opposite polarity. The migration of the (−)s to the top of the rod leaves the right bell (+) and the left bell (−).

The left bell was attached to ground. This bell would be relatively (−) because the local (−) charges in the earth try to get as close as possible to the (+) on the first bell.

So far nothing happens, but if a little metal ball (Franklin's was brass)—suspended by a silk or other insulating thread—is to hang between the two bells, then the ball will at first be attracted to one bell and, immediately after contact (thus acquiring a (+) charge), be repelled. Consequently, it will then be attracted over to the opposite bell—and so on, back and forth—ringing wildly!

Franklin's lightning bells are not a safe "kitchen" experiment for us to try. A Swedish experimenter had actually been killed from lightning discharges in the 1750s; fortunately, Ben Franklin, experimenting during storms, avoided such a fate.

Fig. 1.27 Charges at the bottom of the cloud attract opposite charges from the ground

Also, as mentioned in the introduction, it is not safe to shelter under a tree during a storm. While the discharge, or spark, through the air can take quite jagged paths— and more than one—some possible paths have been sketched in Fig. 1.27.

In this figure negative charges are attracted upwards from tall objects towards the lower (+) of the storm cloud. The discharge, if it occurs, is across the air as well as down multiple pathways inside the tree and down from the tree. One possible path could be via the body of the person standing under a branch.

One of the safest places to be in a lightning storm is inside a car. If the metal of the car should become charged, we know that the charges will get as far away from each other as possible—*i.e.*, to the <u>outside</u> of the metal. This is an example of a "*Faraday cage*"—named after Michael Faraday (see p. 106) who built such a cage to demonstrate that charges reside on the outside of conductors. (The wetness of the tires of the car have no bearing on this, although they do provide a possible discharge path.) We will also see that when the charges suddenly move, the flow is still on the outside due to something called the "*skin effect*" (described in Chap. 2, p. 74).

Thus passengers in metallic cars, planes, and so on are protected from lightning from two distinct points of view.

Finally, although planes are hit many times a year by lightning, the last crash was 40 years ago, because a fuel tank was insufficiently protected. Since that time, aeronautical designers have been extremely careful about the shielding of fuel tanks.

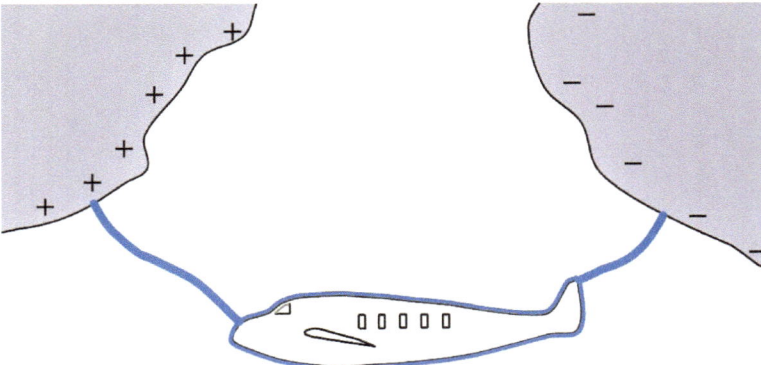

Fig. 1.28 Shows an aircraft flying in the vicinity of storm clouds, with the discharge between the two clouds flowing along the <u>outside</u> of the metal fuselage

Current and Voltage   2

2.1   Water Analogy

In the previous sections we looked at <u>static</u> electric charges, obtained from rubbing combs and other objects, and we looked at the elementary investigations that took place in the 1500s–1800s.

When charges *flow* we call such flow an *electric current*, and a host of interesting things happens. We will find magnetic fields, forces between wires, and even—if we want to muse and ponder—connections to Einstein's special theory of relativity! (This is because who is to say whether the charge is moving relative to us or we are moving past the charge?)

This flow of charges, or coulombs passing every second, is measured in *Amps*, after the absent-minded professor **Andre Marie Ampere** (1775–1836), who once even forgot he'd been invited to dinner with the Emperor Napoleon.

Fig. 2.1 Water analogy, illustrating voltage and current

D. Nightingale and C. Spencer, *A Kitchen Course in Electricity and Magnetism*,
DOI 10.1007/978-3-319-05305-9_2, © Springer International Publishing Switzerland 2015

We make the definition

$$1 \; Coulomb/\text{sec} \; is \; 1 \; Ampere \; (\text{or} \; Amp)$$

and in the water analogy of Fig. 2.1 it would be something like "droplets/second" or "gallons/second" escaping through the little hole near the bottom of the bottle.

This current of water is caused by the pressure, which came from the height of the liquid in the bottle.

Let us briefly review a few of the important experiments concerning electric current, experiments that began in the 1700s.

2.2   Galvani's Frogs' Legs, and Volta's Experiment

In 1780, **Luigi Galvani** (1737–1798), an Italian anatomy lecturer at the University of Bologna, noted that if an electrostatic charge was applied to the muscular legs of dead frogs the legs twitched.

In a similar way, people had sometimes experienced muscle spasms due to accidental shocks when they had been trying to build up large charges with things like Leyden jars. (See Fig. 1.18 on p. 17.)

Galvani—whose name lives on with *galvanized nails*, *galvanometer*, etc.— thought he had found in the frog itself a natural source of electricity, which he might use as a battery. Indeed, there is a fish called the electric ray that does naturally produce electric charge, which paralyzes its victims. The electric eel is another animal that stuns its prey electrically, but it would be somewhat impractical to try to use either of these animals as batteries.

It was shown later by another Italian, the nobleman **Count Alessandro Giuseppe Volta** (1745–1827), that Galvani's frog's leg is not producing electricity, but is merely reacting to the *passage* of electricity, *i.e.*, to *currents*.

The pressing question in Volta's time was as follows: How could these currents be maintained?

Now Count Volta knew that his tongue was a muscle, just like the frog's leg.

Surmising that the muscle of his wet tongue might be more sensitive than a frog's leg, he experimented with different single metals, and touching more than one metal to the tongue caused a definite tingling sensation.

When the metals were removed, the tingling stopped; when put back, the tingling returned. He realized that the *dissimilar metals* somehow caused a continual source of charge.

The next step was to experiment with different metals immersed in acidic liquids and see if the setup could function for an appreciable time.

2.2.1 Tongue Experiment (After *Volta*)

Take one sample of metal, say a copper penny (pre-1982 means copper!), and another type of metal, say a piece of aluminum foil, and place the edge of each quite close together on your tongue.

You will feel the tingling sensation that Volta felt. (If you don't at first it is because the effect is extremely small! Salty water in your mouth will help.)

What is happening is that the slightly acidic saliva on your tongue is adding electrons to each metal, but in different amounts, thus leaving one metal relatively more positive than the other.

Now it was a "mechanical" loss or gain of electrons that happened in our friction experiments, but this time it is to do with the chemical reaction when acids attack metals, and this process is just what we want—if things are to be continuous.

The two different metals may be called (+) and (−), with the one having the larger number of electrons being (−). Naturally, the metals will also be slightly attracted—much too small to observe—but this is no longer where our attention lies.

As usual, the little (−) charges of course want to travel towards the (+)s. Electrons on the electron-rich metal will try to move over to the electron-poor metal continuously, and such a flow of charge goes on and on. Indeed, Volta found the tingling metallic taste all the time he held the metals on his tongue.

The above is thus a simple (but hopelessly weak) battery.

2.3   Experiment: Voltaic Cell

Instead of the mouth, use a beaker or a cup of vinegar—red or white, it doesn't matter. Use two small alligator-type clips to hold a penny (pre-1982) and some aluminum foil in the vinegar, as in Fig. 2.2. (Little alligator clips are easily obtainable from hardware stores; see p. 157.)

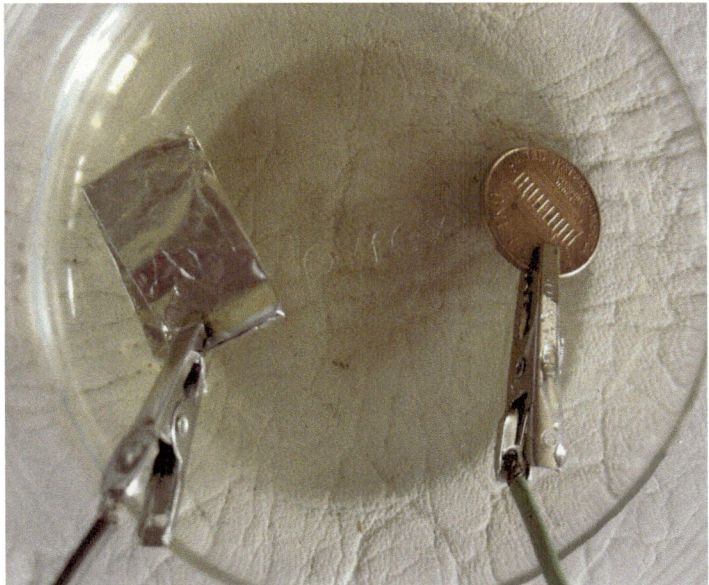

Fig. 2.2 Red vinegar and two different metals battery

Now if we had a sensitive current meter, capable of measuring less than a 1/1000 of an Amp, *i.e.*, 1 milli-Amp, commonly called mA, we would detect a very small current flow between aluminum and copper.

On p. 43 we will make, for educational reasons rather than sensitivity, a crude current meter.

Meanwhile, however, we could purchase an inexpensive multimeter (see p. 158) with a sensitive scale—typically of ~100 <u>micro</u>Amps. (A microAmp is a millionth of an Amp.) The authors found, using such a meter, that the copper coin and sliver of aluminum (as in Fig. 2.2) gave a continuous short-circuit current (short circuit means the + and − are directly connected—here through the meter) of ~45 microAmps or 45/1000 000 of an Amp. To improve matters therefore we should first make a stronger battery.

2.4   Experiment: The Voltaic Pile

The *voltaic pile* is what Volta demonstrated to Napoleon in France (Napoleon had quite an interest in electricity) and, some years later, to Faraday in London.

We will pile up more cells, as Volta did. The aluminum foil can be folded a few times onto itself to make a square roughly the size of a coin, and we sandwich them together with paper towels, or cardboard, again soaked in vinegar. It's helpful to fold the paper towel pieces a few times to make a pad of towel. (Alternatively, pads of vinegar-soaked cardboard could be used.)

Figure 2.3 shows our Voltaic pile, and we say that we have put the cells "in series" (see also p. 64 for things in series).

Fig. 2.3 Voltaic pile. Copper and aluminum, separated by vinegar-soaked paper towel pads

If the kitchen has a metal sink, all that is needed is to put the bottom aluminum "coin" directly on the sink, and then add a folded paper towel pad (which has been dipped in vinegar) on top of the aluminum, then the copper penny, and so on.

Notice that the "vinegared" paper towel provides the central liquid zone, the electrolyte. An electrolyte is a solution in which charged particles such as charged atoms can move about freely. Weak acids, such as dilute sulphuric acid ("battery acid"), vinegar, lemon juice, and the juice in potatoes are good electrolytes, as is common salt solution ("brine"). In fact, any charged particle migrating this way in a fluid may be called an *ion*, extending what we already said on p. 8. (The lightning strike on p. 26 is also an example of ion migration; the air first becomes ionized due to the huge separated charges and then discharges suddenly.)

The vinegar electrolyte lies between the two "coins" of dissimilar metals. Because each cell stands on top of the next cell in the stack, its aluminum coin touches the copper coin of the previous cell, and this contact makes the series connection of each cell to its neighbor.

This is our battery. If a wire is touched onto the top penny and another to the bottom aluminum "coin" (or to the sink if the sink was metal) then we should be able to detect a voltage. Lacking a voltmeter, the tingling in the tongue should be more pronounced[1].

Putting piles *in parallel* (again, see p. 64) will allow more current to pass but yield only the same small voltage. It is possible that such a current may *just* deflect the compass needle of the homemade current meter that we will shortly construct.

2.5   Humphry Davy's Voltaic Pile

After Volta, the young **Humphry Davy (1778–1829)** used a really large Voltaic pile around 1802 to pass large currents through various solutions. (This general process of passing a current through solutions, with an ensuing separation of elements, is called *electrolysis*, and the conducting solutions are, as we said, *electrolytes*.)

In his experiments, this Cornwall-born boy, son of a wood carver and apprentice to an apothecary, not to mention amateur painter and poet, was able to identify, and isolate, the elements potassium and sodium—all by electrolysis! Realizing the importance of what he had done it is recorded that he danced around the lab in joy[2].

Amongst Davy's many achievements he also made an incandescent lamp, using platinum as a filament, but was unable to make it last.

He is known for the discovery of "*laughing gas*" (nitrous oxide) later widely used in dentistry throughout the world as well as the *Davy lamp* for the safety of miners at risk from methane explosions. As a result of his many discoveries in both chemistry and physics, he was knighted, at the age of 34, in 1812. A short while

---

[1] This pile—eight pairs of penny-and-aluminum "cells" in series—yielded a voltage of ~1.5 V and a short-circuit current of about 160 microAmps (or 0.16 milliAmps).

[2] Sacks, Oliver, (2001), Uncle Tungsten, Vintage Books, p.122.

later he married a wealthy widow, and around 1814 he and his wife travelled to France to demonstrate some of his discoveries, with the 22-year-old Michael Faraday as both valet and assistant.

Faraday, of course, would (see p. 106) go on to even greater fame.

2.6   Sidebar Experiment: Electroplating

Before continuing with other ways of obtaining voltages—for example using a potato—we note that Volta's battery may, in principle, be run in reverse.

Recall that we had two different metals in an acid, giving us electricity. Now let us do things backwards, by forcing a current through the acid, using just a common 1.5 V cell. Without going into any chemistry we will find that, with some adjustments, we can cause the material of one metal to be transported across to the other (different) metal! This is called *electroplating*.

We will not use aluminum this time, because aluminum does not work well here, but we will keep copper as the (+) electrode, and the (−) electrode may be for example a kitchen fork, a coin with a surface of some alloy such as nickel, or perhaps a tool from the workshop.

Figure 2.4 shows our basic setup: a glass containing vinegar, plus a little salt to make it conducting. One wire is attached to the copper electrode (here,

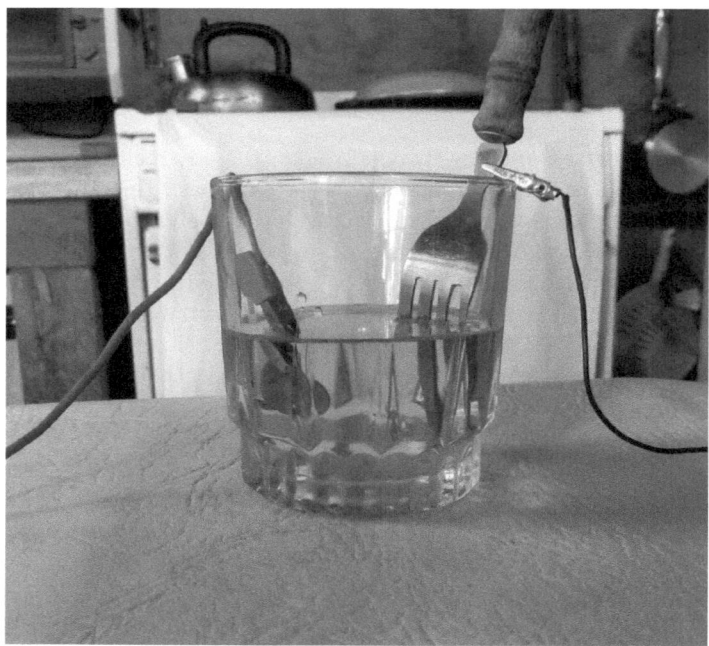

Fig. 2.4 Two electrodes in vinegar. The (+) is a copper penny, and the (−) is the fork or the object to be plated

Fig. 2.5 Copper plating of various objects: A chrome–vanadium wrench, a US quarter (originally a silvery color but actually a nickel alloy), and the tines of a fork

a pre-1982 penny[3] held by a clip), and this is to be connected to the (+) of the household 1.5 V battery. The other wire will connect the object to be plated to the (−) of the battery (battery not shown in Fig. 2.4).

Vinegar is only a weak acid, but it will do the job if we are patient. A stronger acid is muriatic acid, available at hardware stores for cleaning concrete, but it would have to be treated with great caution. Wear eye protection and acid-proof gloves, and read the label.

Our plating results shown in Fig. 2.5 used muriatic acid.

2.7 Experiment: Potato Battery

Going back to our discussion of batteries, we may also make a battery from fruits and/or vegetables, the juices of which are quite rich in ions.

---

[3] We have cut into the penny to make sure that it really is copper. After 1982, US pennies were made of the cheaper metal, zinc.

Fig. 2.6 Making a potato
battery

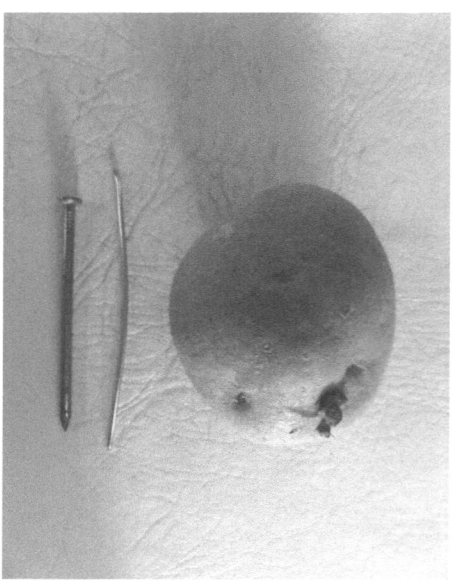

Take two dissimilar metals, *e.g.*, a piece of copper wire and a steel nail as in
Fig. 2.6, and a potato.

Push the two metals into the potato, maybe an inch apart, and clip them to the
leads of the multimeter as shown in Fig. 2.7, choosing either the 20 V range or the
2 V range. Our multimeter here is on its 20 V scale, showing 0.47 V, and so we may
switch to the more sensitive 2 V scale. It will read an extra figure, such as 0.472 V,
although we have no interest in that last digit here.

How did we know to clip the (+) to the copper? In these experiments the copper
electrode is always the positive one, but if we had chosen wrongly, a little minus
sign would have appeared on the digital meter (or the needle of the analog meter
would have wanted to swing backwards).

If we had put three of these potato batteries in series (or, if short of potatoes,
a lemon, a potato, and a tomato as shown in Fig. 2.8) we would have had almost
1.5 V—essentially the voltage of a typical "AA" cell. Unfortunately, such batteries
are not able to pass enough current to do anything useful.

Will potato batteries go on forever? No; ultimately oxides form on the metals,
and slowly the current gets blocked.

2.7.1   What Was Happening

As we saw, different metals in an acid, or what we called an electrolyte, gave us a
voltage. The acid attacked the metals differently, causing one metal to gain more
electrons than the other, and generally it is the copper that becomes more (+).

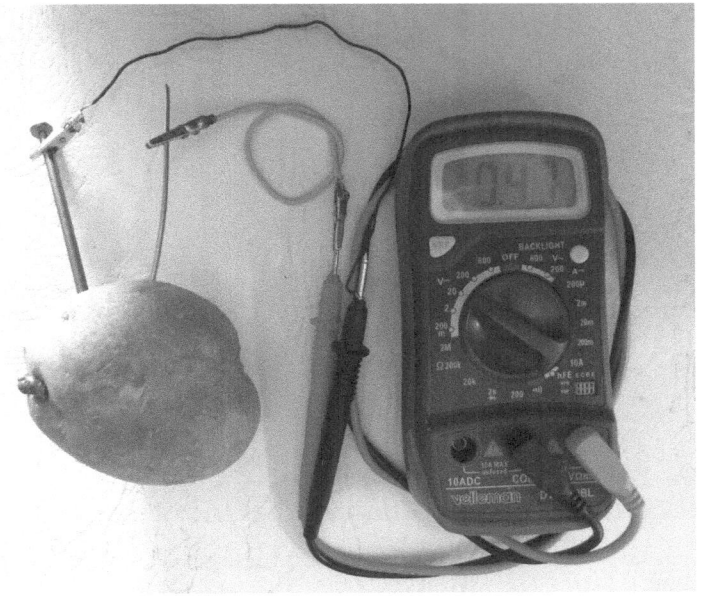

Fig. 2.7   A single-potato battery giving us just under half a Volt

Fig. 2.8   Each fruit yields about half a Volt. Our multimeter here reads 1.47 V

Fig. 2.9 Here we have inserted a dark background in an attempt to show the faint glow from the little red LED

An electrochemist will tell us that it is to do with the relative atomic numbers of the elements. Copper, at number 29, has 2–8–18 in its first three filled shells and the 29th electron is all alone, but steel (i.e., iron) at number 26 has three fewer outer electrons. Briefly, the copper would like to lose that 29th electron, leaving itself (+), but we will not pursue further what the other metals do to fill their shells and become relatively (−) compared to the copper. Instead, in the kitchen, we will see what happens if we try to use this electricity.

If we try to light a bulb with the improved battery of Fig. 2.8 we will be disappointed. However, if we add a fourth vegetable or fruit as in Fig. 2.9—actually here a lemon—we gain another ½ Volt. (The authors' final reading in Fig. 2.9 was actually 1.85 V.)

In principle this would be just enough voltage for an LED—to be discussed later—but there would be too small a current.

To help this we have put the electrodes closer, as well as doubled their contact area, by bending the copper and doubling the nails. The reader may now just be able to discern the little red LED glowing above a black background.

In the next section we clarify further the difference between voltage and current.

2.8   Amps, Volts, Energy, Power

Because newspaper articles commonly confuse "voltage" and "current," not to mention "power" and "energy," which are all different, let's take the water analogy one step further.

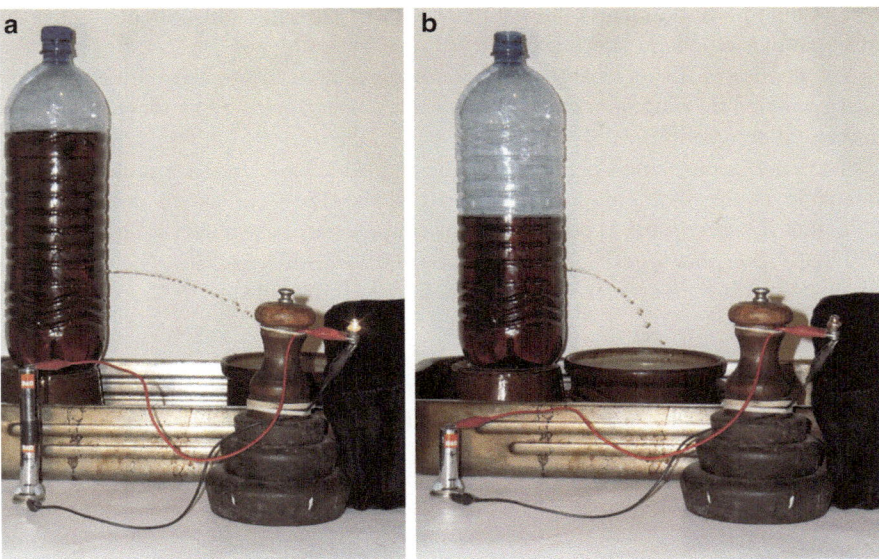

Fig. 2.10 (**a**) Higher level in bottle, and two-cell battery. (**b**) Lower level, and one-cell battery

The diluted coffee in Fig. 2.10a obviously has more *energy* than is shown in Fig. 2.10b—we can see that the two cell battery gets the flow going faster (*i.e.*, larger current, or more electrons/second), and it is thus lighting up the little bulb better than with the lower power in Fig. 2.10b.

We achieved the larger energy by filling the bottle quite high. We say that we increased the "*potential energy*" of the fluid, meaning that it has potential to do greater things, like causing faster currents.

Replace the word "coffee" with "charge." We have more coffee in the left photo or more electric charge.

Suppose we want to drive a little vane or generator in the stream of coffee flowing out of the hole. How much total energy we have available depends BOTH on the pressure (i.e., height) of the water (*potential*, measured in *Volts*) AND the total quantity of charge held in the bottle.

The product of these two is the *potential energy*, because before emerging from the bottle it has, as we said above, potential to do useful work.

So let's remember this:

(*Amount of fluid* × *height*) is *potential energy*

or, same thing,

(*amount of charge* × *potential*) is *potential energy* (or just *energy*, for short).

Energy is measured in *Joules*. (James Prescott Joule was introduced on p. 16.) Using standard international units,

(1 Coulomb × 1 Volt) is called 1 Joule. (Using our same old symbols, this is an example of the general shorthand, q × V = energy.)

As mentioned earlier, many utility companies prefer to use alternative units for energy, like kiloWatt hours (or kWh), rather than Joules (and there's an easy numerical conversion).

The energy consumed per unit of time (*i.e.*, the Joules/s), is a unit of *power*. Its name is the *Watt*, after **James Watt** (1736–1819) of steam engine fame. Note that if energy/time is power, then of course energy is (power × time).

Finally, *potential difference* is to be visualized as the difference between the height of the coffee in each case, in the bottles of Fig. 2.10. It is always measured in *Volts*.

2.9   Experiment: Current Through a Bulb

Because the Voltaic pile is impractical for experiments we resort now to everyday 1.5 V "AA" batteries, such as we find in cameras and flashlights. (We won't yet use the rectangular 9 V batteries; they are more expensive, and it is simply not necessary to have that high a voltage or pressure.)

The following experiment (Fig. 2.11) is trivial and elementary and is the circuit for connecting a bulb to a battery.

In all likelihood there is some kind of a battery-operated flashlight in the household, so let's borrow both the bulb and battery from a flashlight. If you don't have a flashlight, some of the parts that may come in useful are listed in the Appendix on p. 157.

Note: **You cannot get a shock from touching any such low-voltage battery with your hands**.

If you've hooked up the circuit correctly, the lamp will light; and it did not matter which way round the battery was connected. Note also that a switch could be inserted anywhere around the circuit to interrupt the current, but we didn't bother with this because the connection of the final alligator clip serves the same purpose. However, one would not omit a switch in a high-voltage circuit.

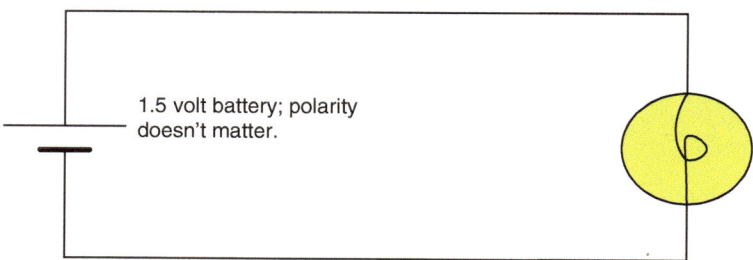

1.5 volt battery; polarity doesn't matter.

Fig. 2.11 Connecting a bulb to a battery. The polarity does not matter

Fig. 2.12 A broken 12 V automobile lamp with two filaments. Without the nitrogen-filled glass bulb the filaments would quickly burn up

2.9.1 What Is Happening

Electrons are flowing from the battery, passing through the slender filament of the incandescent (Latin *incandescere*, to glow) lamp (see Fig. 2.12), and the very thin filament is getting hot—not hot enough to melt—but hot enough to give off light and heat.

The filament was for a long time a stumbling block to **Thomas Alva Edison** (1847–1931) in his invention, or re-invention[4], of the electric lamp in the 1870s, as he tried to get it to last more than a few minutes before burning up. He ultimately discovered that the key was to remove all oxygen, for nothing can burn without oxygen. Although incandescent lamps are dying out now—they are too inefficient, with only about 5 % of the energy being light energy—most incandescent filaments are in a vacuum or in a bulb filled with an inert gas such as nitrogen.

The electrons are of course flowing from (−) to (+). All the currents we deal with, in house and car, are currents of electrons.

---

[4] There had been earlier experiments by Humphry Davy for example and others in different countries. In particular, a patent was issued in 1880 to Joseph Wilson Swan (1828–1914), who had served an apprenticeship in pharmacy, for a low-resistance carbon filament lamp, a lamp that, while it lasted longer than Davy's, still had a limited lifetime because of insufficient vacuum.

If we could get the positive charges to move, that would constitute a current also—but this doesn't happen here, since the (+) charges, which are the much heavier protons anchored in the nuclei of atoms, are not here free to move around.

In the 1700 and earlier 1800s, before it was known that it was the electrons moving, it was conventional to say that current went from (+) to (−). Such "conventional current" is sometimes referred to as the "mathematical current."

2.10   A Fuse

It's trivial to make a fuse from a thin strip of aluminum cooking foil, about 1/4" wide, simply by cutting a notch in it. The notch must leave only a *tiny* thickness of metal (Fig. 2.13).

Connect a battery directly across it—*i.e.*, short-circuit the battery. (Be careful: If you haven't made the "bridge" between the notches <u>very</u> thin, at worst, the battery could quickly overheat and, in extreme circumstances, explode.)

If the bridge between the notches is thin enough the metal will melt.

Meltable-type fuses are found in very old buildings, and under the dashboard (usually) of automobiles.

Today, houses use *circuit breakers*, which achieve the same protection by breaking the flow of current. Although we don't pursue this here, there are two basic ways such circuit breakers may work. An element may heat up and expand and thus break the circuit, or the current through them might trip the circuit by a magnetic device.

Fuses are rated by the value of the current they can pass before burning out or tripping. In the next section we construct a very basic kind of current meter, but we will also find it useful to have access to more sensitive ones, and we have photographed some typical store-bought meters on p. 159.

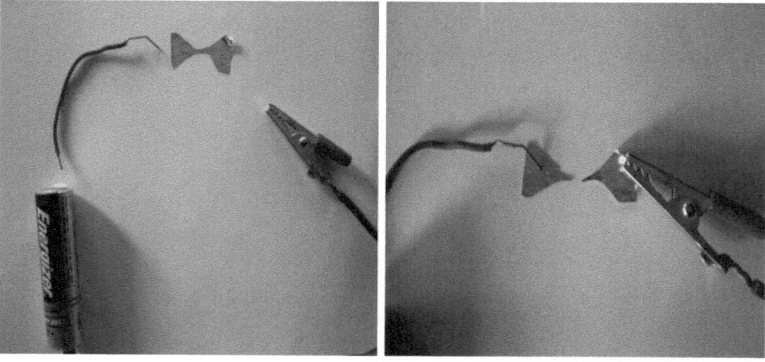

Fig. 2.13 Slice of aluminum foil as a fuse. On the *right* it has obviously burned out

2.11   Making a Current Meter

An elementary "kitchen" meter is shown in Fig. 2.14—even easier to put together than the electroscope we constructed in Chap. 1.

Assuming that there's an ordinary compass in the house, set it on the table, and allow the pointer to come to rest in its natural north–south direction. Make sure that it's away from other electrical devices in the house, such as being too close to the magnet of a radio loudspeaker or the currents associated with a refrigerator motor, etc.

Now take any coil of insulated wire—such as the coil of ordinary connecting wire we used in the experiment of Fig. 1.25—and let it have plenty of turns (the more the merrier). Its ends should be bare; unfortunately it may not be trivial to locate the inner end without having to take a few minutes completely uncoiling and rewinding it!

Now insert this coil of wire into the bulb-and-battery circuit of Fig. 2.11 by placing it in series at any place.

Obviously, the current will then have to also flow through the coil on its way around the whole circuit, as shown in Fig. 2.15.

Make sure that the coil is exactly to one side of the compass needle. If you have no bulb holder some wires can simply touch, as in the photos.

When the circuit is closed (*i.e.*, the wire is touched to the battery) the current flowing through the bulb will also be flowing through the coil—like water traveling through two or three hoses connected one after the other. It *has* to be the same current going through the coil, wire, or battery, because there were no turnoffs or alternative routes.

The bulb will light up just as brightly as before, and you have noticed that the compass needle has deflected from its previous North–South line, as shown by

Fig. 2.14 Coil of insulated wire placed as close as possible to compass (it doesn't matter if it touches the frame of the compass). This is our current meter. Note how we have set it so that the compass originally points north. (Take no notice of the brass marker)

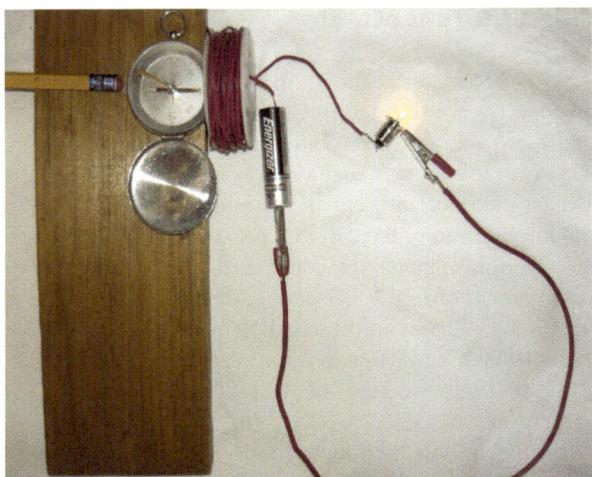

Fig. 2.15 Circuit of Fig. 2.11 modified to include our "ammeter." After connection the compass needle has swung round strongly (direction shown by *pencil direction*). Again brass marker is irrelevant

the pencil. By making a pencil mark one can see how big this deflection is. This is a direct measure, as yet uncalibrated, of the quantity of current flowing around the circuit.

Of course, our homemade meter's coil would have to be placed in <u>exactly</u> the same place relative to the compass each time we make a reading if we are to achieve any kind of accuracy.

With care, this meter could serve in some of our demonstrations, unless greater accuracy is called for.

2.12 Another Way to Get a Voltage: Seebeck Effect

In 1821 **Thomas Johann Seebeck** (1770–1831), a German medical doctor who enjoyed experimenting in physics, discovered that if two dissimilar wires, e.g., iron (or tin) and aluminum, or nickel and copper, were twisted together at their ends and one junction heated while the other junction was kept cool, a current flowed. (He knew this because he noticed the effect on a compass.)

We illustrate the circuit in Fig. 2.16 and a possible setup in Fig. 2.17.

Various food cans are a good kitchen source of dissimilar metals. Our apparatus, shown in Fig. 2.17, consists of two strips: one from an aluminum beer can and the other from a steel tuna fish can. **CAUTION: Not only are these kitchen metals sharp, whatever heat source is used <u>extreme care must be taken</u>. While the meter connections here are the cold ones, in fact the joint (top of photo) must be very hot for a detectable voltage.**

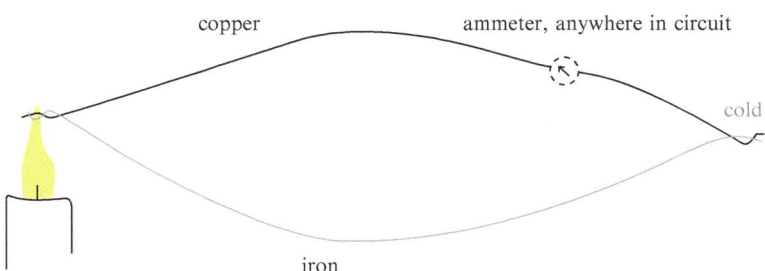

Fig. 2.16 The Seebeck effect. When one junction is hot, and the other cold, current flows round the circuit. (The meter does not have to be inserted as implied above; the *right-hand joint* can equally well be untwisted and the current meter inserted there)

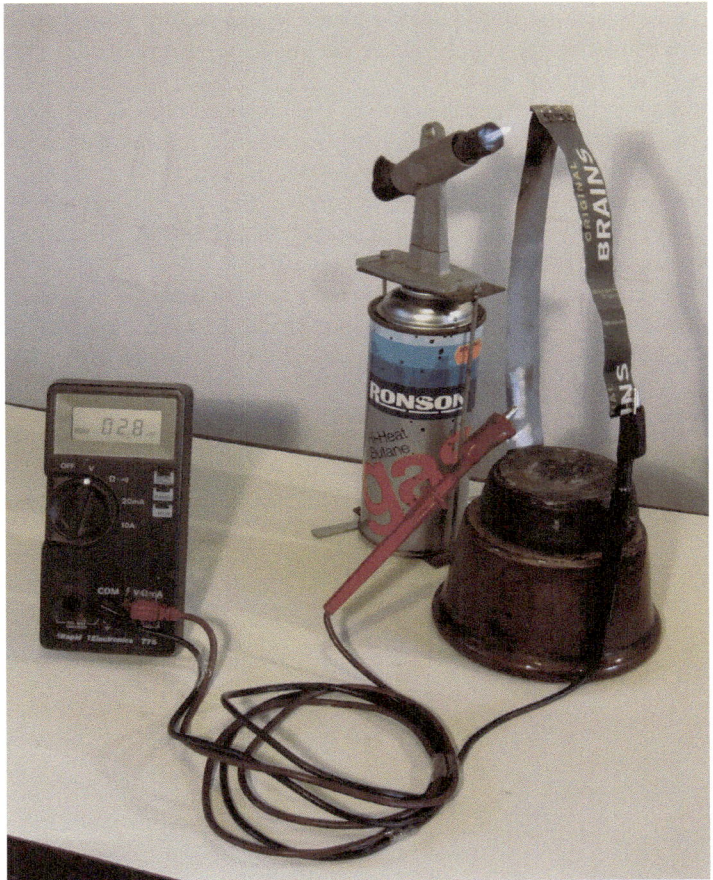

Fig. 2.17 Two dissimilar metal strips, cleaned, joined firmly, and heated (here by a blowtorch). To get even a small reading on the voltmeter one joint must be very hot—**so extreme care should be taken**. The apparatus could alternatively be set up with the strips approximately horizontal, and then the hot joint could be on a gas or an electric ring

Aluminum and steel (96 % iron) make a good Seebeck pair.

Using an old pair of scissors, a 1″ wide ribbon of Al was cut from the beer can, and a similar ribbon of steel from the tuna fish can—both about 6″ long. Each end of each ribbon was abraded with emery paper to remove the thin plastic internal coating which stops the metal reacting with the contents. One end of each was put together, abraded areas in contact, and then the end of the double strip was crimped several times, like an old-fashioned toothpaste tube, and hammered to make sure that there was a firm mechanical and electrical contact. (With such a good electrical connection the resistance would be a small fraction of an Ohm. For the definition of an Ohm see pp. 50–52.)

From Internet data this combination is expected to give about 1.5 mV when the cold ends are at 0 °C and the hot junction at 100 °C.

One of the analog meters (see p. 159) has a 100 mV setting, and a typical digital meter has a 200 mV setting, so the Seebeck voltage should be just visible on either. However, the dissipation of the analog meter would require more current than a small Seebeck junction can supply, hence our above procedure making sure that we have a physically large junction.

It turns out that the steel end is positive. The crocodile clips of the meter needed to be clipped on to the free ends of the strips where the coating has been removed (as directed above).

As can be seen on the digital meter in the photo (Fig. 2.17) the voltage is 2.8 mV, which accords with typical data for such a junction, and our junction was considerably more than 100 °C hotter than the cold ends.

Blowlamps are available in local hardware stores, if kitchen gas or electric rings are not used. For the latter, the setup can be rotated through a right angle to heat

Fig. 2.18 Fan is powered ONLY by heat from the stove, using the Seebeck effect

from the stove or range, and it's OK to have one strip touching metal parts of the range, as long as both don't!

Although the Seebeck effect is very small it has been much improved upon since his time and can nowadays do things like drive a small fan on top of a wood-burning stove or other hot surface, (*e.g.*, the Canadian-made Ecofan, Fig. 2.18).

Seebeck thermocouples can not only power little motors but could also serve, of course, as general indicators of any very high temperatures.

2.13   Peltier Effect

Jean-Claude Peltier (1785–1845), a French watchmaker who turned to physics, discovered about a dozen years after Seebeck's discovery that instead of having a hot junction and a cold junction causing a current, actually passing a current will cause one of the junctions to get cold and the other to get hot.

It is too difficult to make any kind of an effective Peltier device at home, because the temperature differences produced will be unsatisfactorily small. However, mini-refrigerators that can keep a drink cool in an automobile, etc. are available on the Internet, and these illustrate Peltier's original experiment.

2.14   Yet Another Way to Get a Voltage: Piezoelectricity

The word "piezo" comes from the Greek word for "to squash" or "to compress."

If one squashes a crystal of Rochelle salt—or quartz or cane sugar—a tiny voltage will appear across the crystal. (Rochelle salt should not be confused with rock salt—halite—which is no good for the piezo effect.)

The piezo-electric effect was investigated in the 1880s by the brothers **Pierre Curie** (1859–1906) and **Jacques Curie** (1856–1941) in Paris.

Figure 2.19 shows the basic idea.

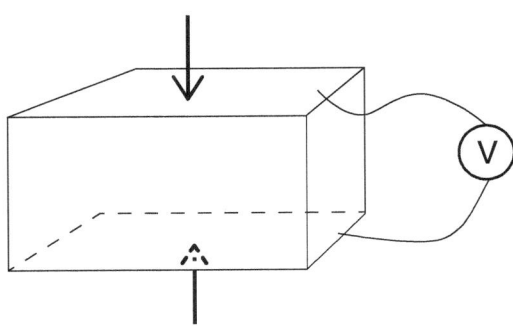

Fig. 2.19 Pressure on certain crystals, such as quartz or cane sugar, separates charges and produces a voltage

Fig. 2.20 Cane sugar and a sample of quartz

The converse effect—that a voltage would distort the crystal—was predicted in 1881 by **Gabriel Lippmann** (1845–1921)[5] and was immediately verified also by the Curie brothers.

By squashing either of the above crystals, the experimenter might think that a voltage could be detected, but this is not so easy—and it leaves one in total admiration of the experimental skills of the Curie brothers. However, a "transducer" (see footnote on p. 136) utilizes a piezo crystal and our Fig. 2.21 shows what happens when pressure is applied with a cocktail stick to a transducer. A twitch of about 1 mV has registered momentarily on the analog meter, and when the pressure was released the meter showed a twitch the other way, of about (−1) mV.

Many devices use the piezo effect, such as microphones (where the rapidly changing air pressure—or force—on the crystal causes corresponding changing voltages) and where changing voltages will cause a corresponding change in size (thus acting like a loudspeaker). We have used exactly this on p. 136.

---

[5] Lippmann was a professor at the Sorbonne (and later Nobel Prize winner for obtaining colors photographically by means of interference methods). He died aboard the steamer *SS France* at the age of 75 while returning from Canada to France.

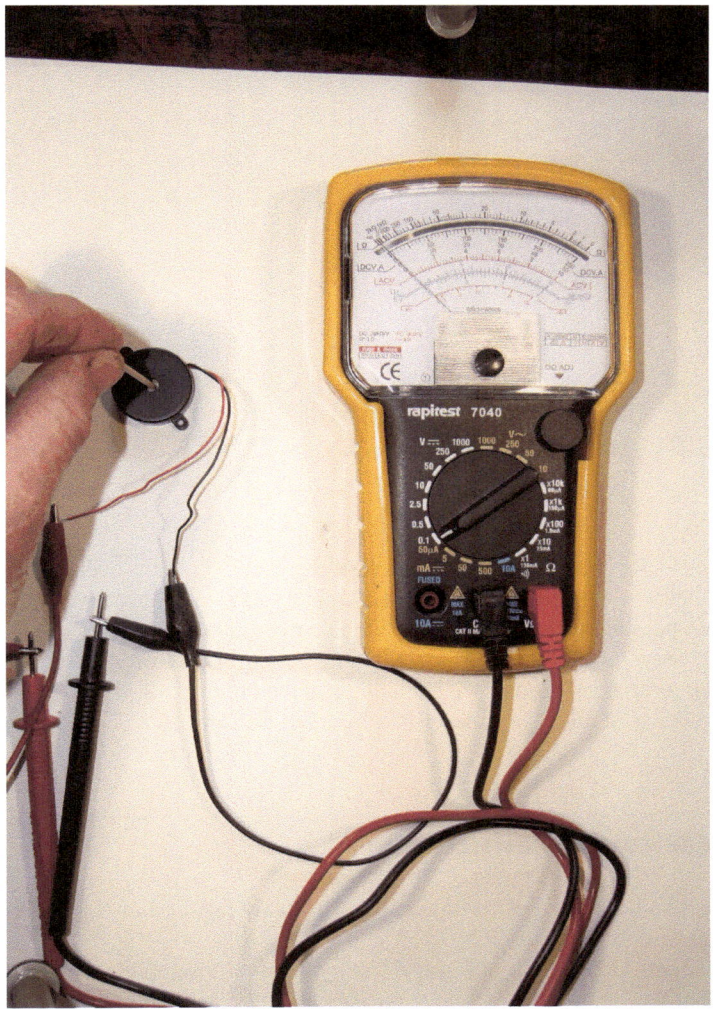

Fig. 2.21 Flicking a piezo sample

2.15   L.E.D.s vs. Bulbs

L.E.D.s, or light-emitting diodes, are increasingly used nowadays instead of bulbs, and they have one profound advantage: they use very little current for the same amount of light. The photo on the next page shows an assortment bought in a packet. So, instead of the little incandescent light bulb of Fig. 2.11, couldn't we just put a battery across one and see it light up?

Fig. 2.22 An assortment of LEDs

Emphatically NOT YET!!! (**Please wait for p. 55.**) There is every chance it will burn out, because the L.E.D. (hereafter LED) is a solid-state device, very sensitive to the current passing through it, and it will also need the battery connected with the correct polarity—which the ordinary bulb does not. All LEDs must have their current limited (by a resistor)—usually to something of the order of ~20 mA.

The working of an LED is described in detail in Chap. 4.

2.16 Concept of Resistance

Some metals allow charges to flow easily—we saw this earlier—while others are not so good (see table on p. 124).

The filament of the bulb in Fig. 2.11 allows only a limited current. We say that the filament has a certain *resistance*, measured in *Ohms*, after **Georg Simon Ohm** (1787–1854), a German school teacher of mathematics and son of a self-educated locksmith who managed to teach his children all he knew about mathematics, physics, chemistry, and philosophy.

Just as a bigger pressure of water forces more gallons/second to flow, as in Fig. 2.10, so a higher voltage will cause a higher current. If there's too much pressure, the water pipe may burst; and analogously, if there's too much voltage too many electrons will flow and the filament will melt.

Also, in the same way that a narrow pipe *resists* the flow of current more than a fat pipe, so a narrow wire has more *resistance* (Ohms) than a fat wire. Further, if we were to experiment with a hose, a long hose will resist the flow more than a short one. We summarize this by saying that resistance to flow is directly proportional to length but inversely proportional to cross-sectional area.

Fig. 2.23 100 Ohm resistors—five low power ("wattage") and one high wattage

A good example of the above is the cable from an automobile battery, which is fat and short. The cable has to be able to pass the <u>huge</u> flow of electrons taken by a starter motor—maybe 100 Amps.

2.17 Ohm's Law

Because we have that "the bigger the pressure, the bigger the current," we can write that pressure is proportional to current or

$$\text{Pressure} = \text{current} \times a\ constant.$$

Georg Ohm called this constant the "*resistance*" of his circuit and enunciated his law in 1826. The law is used every day by people working with electricity (provided there <u>is</u> a direct proportionality between V and I—not the case with LEDs for example). We write

$$\textbf{Pressure} = \textbf{current} \times \textbf{resistance}$$

or

$$\textbf{voltage} = \textbf{current} \times \textbf{resistance}$$

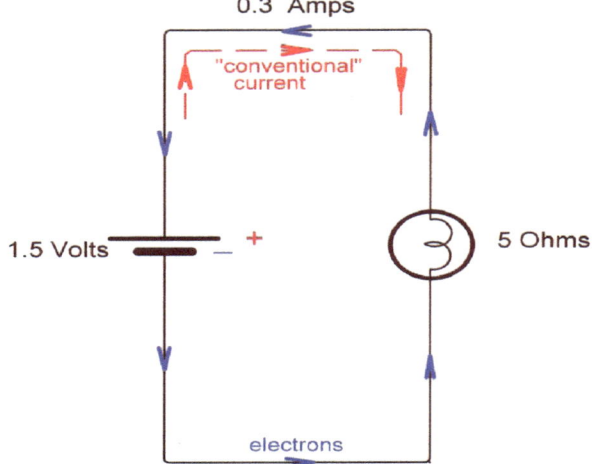

Fig. 2.24 Illustration of Ohm's law, here yielding R = 5 Ohms. Note: When a bulb is "cold," its resistance is initially small. When it reaches its final steady temperature, its resistance settles at a somewhat higher value. It is generally true that resistance goes up with temperature (there are exceptions, and there is no obvious water analogy here). This accounts for the initial brief surge of current when circuits are first switched on

or[6], using a shorthand,

$$V = I \times R.$$

Example:

The voltage of our battery in Fig. 2.11 is 1.5 V.

The current that will pass through the bulb (it's on the package!) is ~ 300 mA (0.3 A) IF used with the 1.5 V battery.

From Ohm's law, 1.5 V = 0.3 Amps × the resistance of the filament.

From simple arithmetic, this means that the resistance (R) of the bulb must be 5 Ohms.

We should be aware that there is also some small *internal resistance*, of about 1/10 of an Ohm, in the battery itself, which can be neglected here. In an automobile battery it is even smaller—about 1/1000 of an Ohm. Figure 2.24 shows the electron flow and the values of voltage, current, and resistance.

Although we have shown a coiled filament in Fig. 2.24, the standard symbol for a resistor is given together with other common symbols on p. 63.

---

[6] We said "no formulae"; these are just shorthands, like acronyms. So one might be happier with V = CR, but the world always uses I for "current."

Fig. 2.25 A representation
of Ohm's law. (The graph for
the resistance of the filament
of an actual incandescent
lamp approximates this,
departing slightly from
linearity, particularly at the
high-current end where
things get hot)

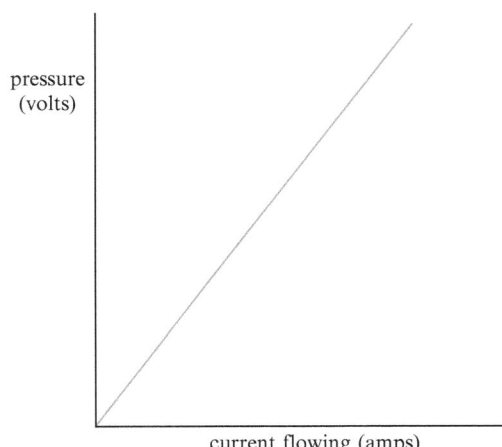

pressure
(volts)

current flowing (amps)

2.17.1 A Graph for Ohm's Law

Notice that the bigger the pressure (vertical axis) then the bigger the current
(horizontal axis). We call this straight-line graph a "linear" graph. Another word
for this linear current–voltage graph is "Ohmic," and a light bulb is approximately
Ohmic. However, we'll see later that things like LEDs are definitely <u>not</u> linear—
they do not have nice straight-line graphs relating current and voltage. We call
them, rather obviously, "non-Ohmic".

For comparison, the graph for a non-Ohmic device is on p. 130.

2.17.2 Experiment: Resistance of a Household Bulb

Replace the flashlight bulb of Fig. 2.15 on p. 44 with a household incandescent bulb
(unscrew one from a fixture, assuming that you still have one!). It doesn't much
matter whether it's a 50, 75, 100, or 150 Watt. We will find that this ordinary house
bulb has a considerably higher resistance than the flashlight bulb.

When we touch the battery leads onto the terminals of the big household bulb
(instead of the small bulb) the homemade compass meter will read less—considerably
less. If there isn't a different reading at all with a 60 Watt bulb, try one which has a
rated (*i.e.*, printed on the glass of the bulb) power of 150 Watts or so.

In any case, whichever household incandescent bulb is used, that bulb's resistance is indeed higher than that of the little flashlight bulb.

2.17.3 What Was Happening

The *power* of the heat and light radiated from a bulb depends not only on the
voltage but also on the number of electrons rushing through the filament every

Fig. 2.26 Here, less current flows than in Fig. 2.15, and the pencil shows the compass needle's new position. (In this photo it happens to coincide with the brass "marker," which is irrelevant)

second, *i.e.*, the current. With the flashlight bulb, there were a lot of electrons able to pass through the low-resistance filament, and of course we only needed a small electrical pressure (the voltage was only 1.5 V) to achieve this.

With the big household bulbs we observed that *less* current was passing, because the compass needle wasn't deflected as much as before. In order to achieve the higher power output of light and heat, something had to be different. What would that be?

First, notice that the pressure (*i.e.*, household voltage, ~120 V in America for example) is nearly 100 times greater than the battery voltage. But this kind of pressure, surely, would cause a huge average current to flow?

No, because the bulb was designed with a bigger resistance. The makers <u>had</u> to provide a considerably higher resistance filament than the one in the flashlight bulb, if they wanted to keep the current flow small. After all, that's what the homeowner pays for—the current (or charge/second)—multiplied by the volts and the total time, with the result all registered by the meter outside.

2.18   Equivalent Definition of Power

As mentioned earlier, power (*Watts*) was (*charge* × *voltage*) divided by *time*. Since we learned also that (charge/time) is current, then power can equivalently be written as (*Volts* × *Amps*)—or again in <u>shorthand</u> (!)

$$\textbf{Power} \ (in \ \textit{Watts}) = \mathbf{V} \times \mathbf{I},$$

a definition that will be useful to remember as general knowledge.

Equivalently, since we already know Ohm's law, we needn't use voltage at all, and we can just write the above as power $= (\textit{current} \times \textit{resistance}) \times \textit{current} = (I \times R) \times I$.

Because "current" appears twice this is often referred to as "I squared R" heating or sometimes, as alluded to on p. 16, "Joule" heating.

2.19 Lighting the LED

We mentioned that an LED will be destroyed if we try to pass too much current through it.

In the circuit of Fig. 2.27 the voltage from the two "AA"s in series is $1.5 + 1.5 = 3$, and the red LED at the top left is giving off light. The current is safely limited by the two 100 Ohm resistors in parallel (small, twisted together, top right), and this is equivalent to one 50 Ohm resistor. (The current divides into two equal paths.)

We mentioned that the LED, because it is a diode, is certainly NOT Ohmic, so its resistance is not a fixed value but depends on how much current we are pushing through it. (Again, see Chap. 4, p. 130, for the actual current–voltage graph.) These small LEDs typically need roughly 2 or 3 V across them in order to get something like 15–25 mA flowing, and such information is often given on the packet.

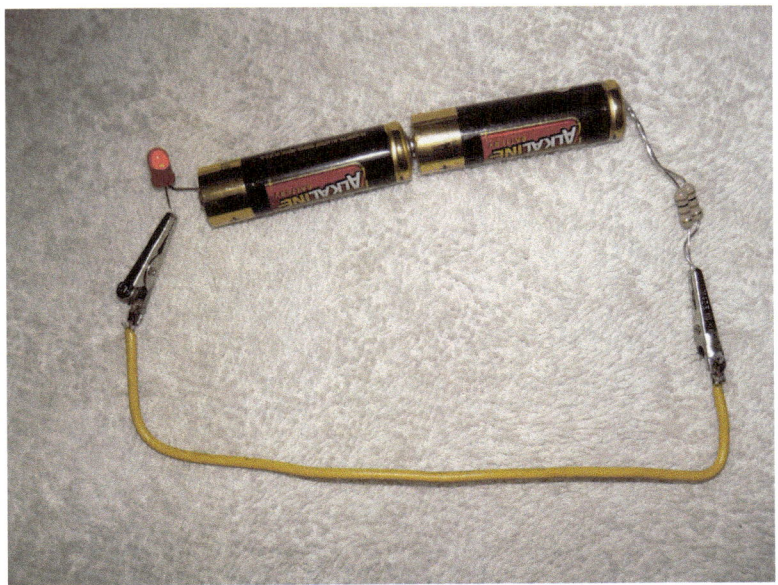

Fig. 2.27 Tiny red LED at *top left*; 50 Ohms at *top right*

So what is the resistance of the little diode (in Fig. 2.27) at this value of current?

We already have a 50 Ohm resistor in series in the circuit. The total resistance "seen" by the battery is therefore ~(50 + R) Ohms, where R is the diode resistance, again at these values.

In this setup the authors measured (this time with a multimeter) approximately 15 mA passing round the circuit. Thus, from Ohm's law, 0.015 Amps = Volts/resistance = 3 Volts/[50 + R] which implies that R at these values = 150 Ohms.

So our LED lights up! (If it didn't, it was the wrong way round; by convention, LED manufacturers make the (−) side flat.)

Another way to look at the above is to refer to *voltage drops*. The voltage drop across the 50 Ohm resistor is I multiplied by R (from Ohm's law, and the pure resistance is Ohmic) or $0.015 \times 50 = 0.75$ V, and the voltage drop across the LED is 0.015×150, *i.e.*, 2.25 V. These two voltages add up to 3 V, and this is an illustration of one of the two basic laws given by **Gustav Robert Kirchhoff** (1824–1887) of Konigsberg, Prussia, later a professor of physics at Heidelberg and Berlin. (Kirchhoff was a friend of Bunsen, of Bunsen burner fame, and Kirchhoff is also well known for having calculated in his early 30s that the speed of a signal along a perfectly conducting wire must approach what is now known to be the speed of light.)[7]

We don't consider Kirchhoff's laws mathematically in the kitchen, but they are easy to visualize. At a kitchen sink for example the hot water and the cold water mix at a junction and then come out of one pipe into the sink. Kirchhoff's junction law, usually referred to as his first law, says that at any joint (or junction) the total current of water flowing in is balanced by the total current of water flowing out.

Kirchhoff's second law has already been illustrated in the paragraph before the above, where we added up all the "voltage drops" (pressure changes) around a loop, and they summed algebraically to zero—*i.e.*, they "balanced." An analogy might be visualizing steps of water pressure in a multi-storey building—the pressure at the ground floor apartment might be 60 (units), the pressure at the apartment above it might be 50, at the next level might be 40, and so on, but the pressure *differences* (10 each) between each floors must add up to the *total* pressure difference between top and bottom.

The exact expression of Kirchhoff's two laws is given in Appendix F.

2.20   The Solar Cell: A (Part-Time) Battery

Instead of battery energy we can often use the energy of the sun.

When we studied electrostatics, it was friction that dislodged the electrons, and with batteries it was chemical reactions. We now know of two other possible ways

---

[7] The speed of light had been measured roughly by various experimenters in the eighteenth century but was not measured accurately until the late 1800s—notably by the American Nobel Prize winner A. A. Michelson, whose famous experiments took place about the time Kirchhoff died.

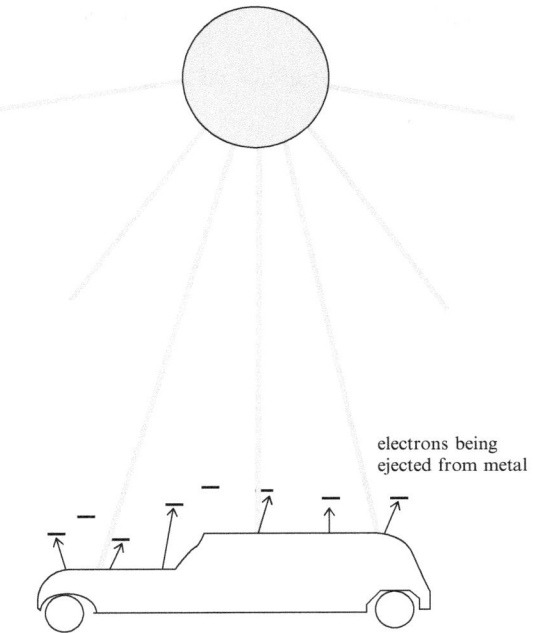

Fig. 2.28 The photoelectric effect: Radiation causing the emission of electrons. Note that this is NOT the method used with everyday photovoltaic cells

of dislodging electrons: one is by the photoelectric effect, known to Heinrich Hertz in the 1800s and explained by the young Einstein in 1905, and the other is by the use of semiconductors using (especially) silicon.

When radiation falls on various bare metals, *electrons are given off*. This is the *photoelectric effect*, shown in an ultra-simplistic way in Fig. 2.28 (assuming that the car is unpainted). **This method is NOT used to provide electricity from the sun—it's just not large enough**.

However, in 1939 the American physicist **Russell Shoemaker Ohl** (1898–1987) of Bell Labs was experimenting with a piece of silicon which had a crack in it. Specifically, he was measuring the sample's resistance. To make a long story short this cracked sample of silicon happened to have slightly different contaminants (or impurities) on each side of the crack and thus slightly different electron densities; it was a primitive "*pn*" junction. We will discuss the basic *pn* junction in detail in Chap. 4.

Now Ohl thought that, because of the different electron densities, a small current might therefore flow across the crack. But it didn't.

However, something else happened. He noticed that, when sunlight fell on his silicon sample, electrons did flow across the crack! He had stumbled across the first "solar cell" or "*photodiode*."[8] The ensuing patent he obtained was Patent #2402662, "Light Sensitive Device."

Modern *photovoltaic* (*pV*) arrays are based on this astonishing discovery.

2.21   More on pV Cells (Solar Cells or Photodiodes)

After some years, research on solar cells became more widespread. The oil embargo of the 1970s together with later oil crises forced many countries to increase efforts on alternative sources of energy. Production of semiconductor solar cells began to be increased mightily, and the price of solar cells in terms of power dropped from ~$300/Watt in the 1950s to <$4/Watt by 2012.

Conversion of light energy into electrical energy via *pV* cells is now common, and one sees small weather-detecting devices in parks and other remote places being powered by the sun. *pV* cells are increasingly seen on the roofs of buildings, legally pumping electricity back into the public grid.

We can obtain standard solar cells quite easily from stores, and here are a few points to note:

On cloudy days we obviously get only a small voltage from the ambient light, but in full sunlight we get a particular voltage (stated on the solar cell) which is able to drive a current around a circuit, depending on the total resistance of that circuit.

Figure 2.29 shows a bunch of broken solar cells (from a $10 grab bag) glued onto a piece of plywood and soldered together serially to provide enough power to charge a 1.5 V "AA" battery. (It must be a rechargeable type; it is impossible to charge ordinary batteries.)

Let's do an arithmetic example concerning a solar cell, again using only the definition of power.

Example: If a $5 solar cell can deliver 2 Volt and 1/10 Amp of short-circuit current (if we just short it out) in full sunlight, what is the electrical power being generated? What is the cost/Watt?

Answer: Power = 2 Volts × 0.1 Amps = 0.2 Watts (*i.e.*, 1/5 Watt). This corresponds to about $25 per Watt.

2.21.1 Actual Solar Cells from the Stores

Radio Shack sells a very small solar cell for about $5, which yields, in full sunlight, a "pressure" or a voltage of ~0.45 V. It can deliver a short-circuit current of as much as 0.3 Amps. (If it is connected to a load the charging current will be less than 0.3 Amps, depending on the resistance of the *external* circuit, as we saw earlier.)

---

[8] Photodiode is a loosely used term. It may be a small device that produces a small current, as in a bar code reader, a larger device (solar cell) that produces more current, or a device that, when current passes, produces light (LED). However, any way you look at it, it is a *pn* junction.

Fig. 2.29 Some "reject" cracked solar cells, glued onto a spare piece of plywood and soldered in series, adding up to ~2 V—enough to charge a rechargeable "AA" battery. This little set of broken pV cells has been trickle charging AA camera batteries for many years, one or two at a time (although actually four batteries appear in this photo)

Fig. 2.30 A 14 V solar panel. (In the *background* are rechargeable "AA"s plus an old ammeter)

However, we can find cheaper " $/Watt" if we go to bigger cells, from online catalogs such as "www.herbach.com" and/or "www.excite.com" (which carries the Edmunds Scientific catalog).

This $39 solar panel in Fig. 2.30, bought from a catalog, has a stated "full sunlight" voltage of about 14.4 V (although 17 V was measured on a particularly sunny day).

The solar cells and panels illustrated in the preceding photos have been in use by one of the authors for over 20 years, showing no degradation other than cobwebs or being knocked over by cats. Cracks are rarely a problem.

2.21.2 Note on Rechargeable Batteries: NiCad, NiMH, Li-Ion

Alkaline batteries are, as mentioned, <u>not</u> rechargeable. Below are some properties of commonly available rechargeable batteries:

2.21.3 Nickel Cadmium (NiCad)

The basic NiCad cell has a voltage of about 1.2 V and may be used where previously ordinary 1.5 V cells were operating cameras, etc. Their internal resistance is smaller than that of regular alkaline batteries.

NiCads sometimes suffer from a "memory" defect where they tend to only charge up to some "remembered" previous value.

Cadmium is a toxic metal, as mercury is, and is thus bad environmentally—one does not need these in landfills.

2.21.4 Nickel-Metal Hydride (NiMH)

The basic output voltage for these is still about 1.2 V, and many of the Toyota Prius models use six cells in series (*i.e.*, 7.2 V), in a bank of 28, amounting to a final DC voltage of 201.6 V. Other hybrid vehicles that use NiMH batteries are the Honda Insight (100.8 Volts) and the Ford Fusion (275 V).

NiMH batteries have almost double the capacity (*i.e.*, the amount of charge they can hold) of that of the NiCads.

They have a lesser "memory problem," although it is still there.

2.21.5 Lithium-Ion (Li-Ion)

Lithium is the lightest of all metals and has only three electrons—two in its first shell and only one in the next shell. The metal is unstable, but in ionized form—*i.e.*, having lost its outer electron—it is safe and usable.

Lithium-ion batteries are light in weight and have no "memory" effect. The voltage of the basic cell is between 3 and 4 V.

These batteries do not contain toxic materials such as cadmium or mercury.

In extreme heat all the above rechargeable batteries can explode. Related to this, they should never be short-circuited, for more than a fraction of a second. (Note that we <u>will</u> very briefly "tap" a regular <u>alkaline</u> battery in the Magnetism section, Chap. 3).

A drawback to all rechargeable batteries compared to alkaline ones is that they cannot hold a charge for a long time, whereas alkaline batteries can stay on the shelf for many years.

People using digital cameras may have noticed that a perfectly new regular alkaline "AA" very quickly causes a message "battery low." In fact, the battery is not low and could be used for many other purposes—flashlights, etc. The problem is the large current needed for the capacitor behind the flash to recharge—perhaps 0.5 Amps, and, because of the relatively high internal resistance of those "AA"s, there is quite a voltage drop as they charge this capacitor. Thus the misleading camera message is "battery low."

Interestingly, the rechargeables, even with their lower output voltage of 1.2 V, can manage this current on account of their much smaller internal resistance. However, nothing is perfect—and because the rechargeables don't hold a lot of total charge compared to the alkalines, it remains a toss-up as to which type to use in any particular application.

2.22   A Charging Circuit, and a Difficulty

What is to stop our newly charged battery from discharging back through the solar cell[9] after the sun goes down? Such a possible "leaking back" is analogous to having pumped up a tire, or a pressure tank, without a *check valve*, and the electrical analog for this check valve is the diode. Its symbol in any circuit is fairly obviously a black arrow (Fig. 2.32).

The diode has a very low resistance in one direction (the *forward* resistance) and a very high resistance in the *reverse* direction. The diode with the smallest forward resistance is actually a Schottky diode, named after the German physicist **Walter H. Schottky** (1886–1976), who did much theoretical work on early semiconductors. Its symbol is similar to the plain diode symbol but with an "S" instead of a line, as in Fig. 2.32.

The circuit of Fig. 2.31 has its diode included (either plain or, with less of a voltage drop, a Schottky diode)—and it may be inserted anywhere around the loop. Our circuit diagram also has an optional current meter, should we wish to check that the sun is actually charging the battery.

Alternatively, we could learn this merely with an LED.

A device to tell us how full a battery is after charging in the sun would be a very useful thing. In the case of a metal water tank we can of course not observe (directly) how full it is. In the case of an automobile's gas tank we might have a float giving a signal on the dash. In principle there would be ways to measure the

---

[9] The reader may object that solar cells are generally diodes anyway, so there can be no leaking back. However, if some of the cells are damaged, or for some reason nonfunctional, an external diode will be an insurance.

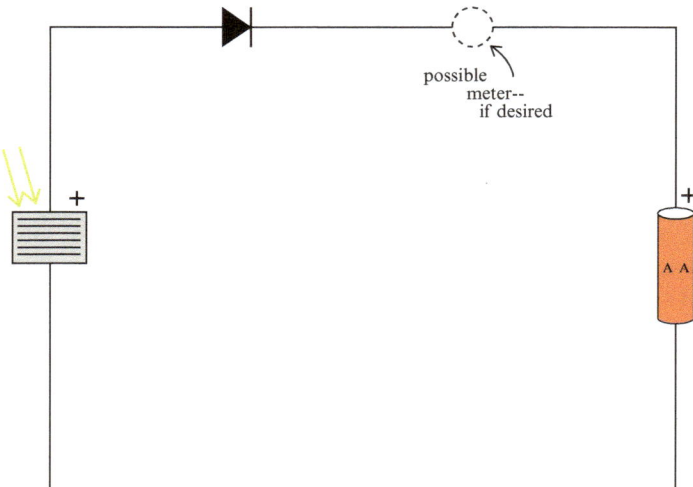

Fig. 2.31 Solar cell charging a battery

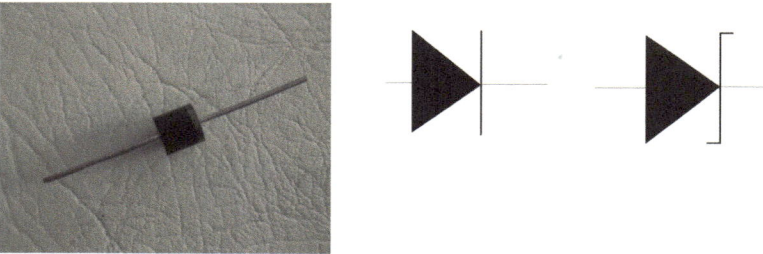

Fig. 2.32 Photograph of a plain diode, with its obvious symbol, followed by the symbol for a Schottky diode

total charge in the battery—such as by seeing how far the leaves diverge on an electroscope, but in reality it's not that simple in the kitchen.

For comparison, computers have a nice little "bar graph" for their state of battery charge, but such circuitry is more complicated than we wish to go into here.

2.23 Brief History of Electrical Diodes

The idea of a diode goes back to at least the early 1900s, when *galena*, or lead sulphide, was shown to conduct currents more easily one way than the other.

Other substances, such as carborundum, copper pyrite, and iron pyrite (fool's gold), were also observed by experimenters to have this asymmetry—in other words they were natural *rectifiers* (see p. 72, AC/DC).

In the early days of radio, where it was necessary to "rectify" modulated radio waves, galena was used in the so-called *cat's whisker* of crystal sets. In Chap. 3 we show how it's possible to actually make a cat's whisker type of radio such as was common in the 1920s. The "whisker" itself was merely the thin wire making a good contact with the galena.

In the last half of the twentieth century electronic valves (*i.e.*, vacuum tubes) were used in all radios and TVs, and these tubes allowed current to flow only one way (which of course is why they were called *valves*). Electrons came from a cathode that was heated by a hot filament, and they then flowed up to the positive anode (or *plate*); they could not flow the other way. Some of the tubes also had metal grids located between cathode and anode which could decrease and/or amplify the electron currents. Such glass-enclosed devices—triodes and pentodes—are still used today, especially in high-quality power amplifiers.

2.24 More Symbols

The devices represented by the symbols shown in Figs. 2.32 and 2.33 are based on the *pn* junction, to be discussed in Chap. 4.

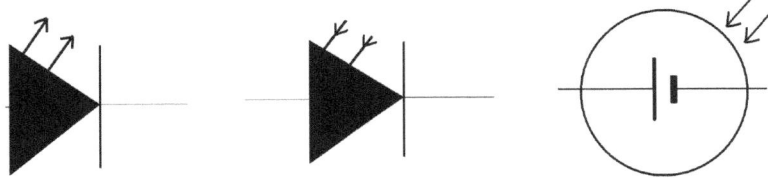

Fig. 2.33 Symbols for, respectively, LED, photodiode, and solar or pV cell

Other common symbols used in electricity are given in Fig. 2.34:

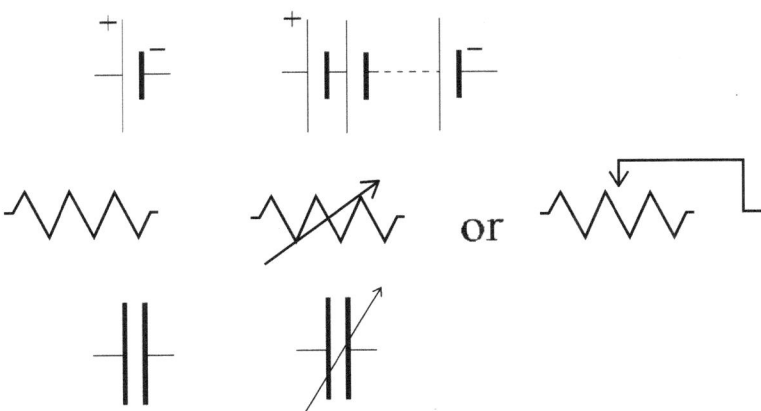

Fig. 2.34 *Top row*: Single cell, battery of cells; *middle row*: resistor, variable resistors (either/or); *bottom row*: capacitor, variable capacitor

2.24.1 Comment on the Various Uses of LEDs:

In bar code readers (see p. 163), the pattern of bars is illuminated with light from an LED and the reflected image is scanned across the sensitive area of a photodiode. When a black bar image falls on the photodiode it blocks the current in a circuit, and the bar is registered in the reader memory. When a blank space image falls on the photodiode, it allows a current to pass and registers a blank in the memory. With a fixed scanner (supermarket or library) an optical system sweeps the bar code image across the photodiode. With a handheld scanner the operator moves the scanner along the pattern.

In the solar cell (or pV cell) a voltage appears across the terminals as soon as it is illuminated (as may be seen with any meter).

Interestingly, it is not commonly known that we could, in principle, use an LED as a (quite inefficient) solar cell. If we were to place a bank of LEDs in the sunlight we would get a minuscule current.

2.25   Series and Parallel: Water Analogy

As mentioned before, in electricity we can connect things "in series" or "in parallel." In Fig. 2.35, two 1.5 V batteries are connected first *in parallel* and then *in series*:

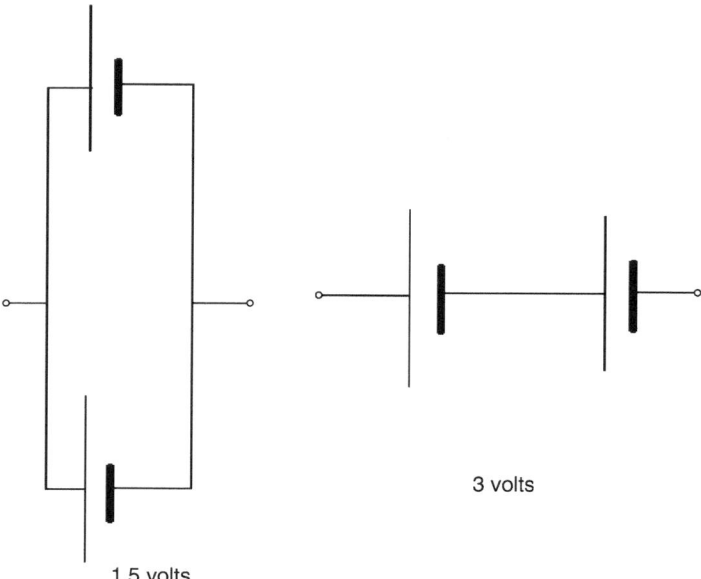

Fig. 2.35 Batteries in parallel and in series

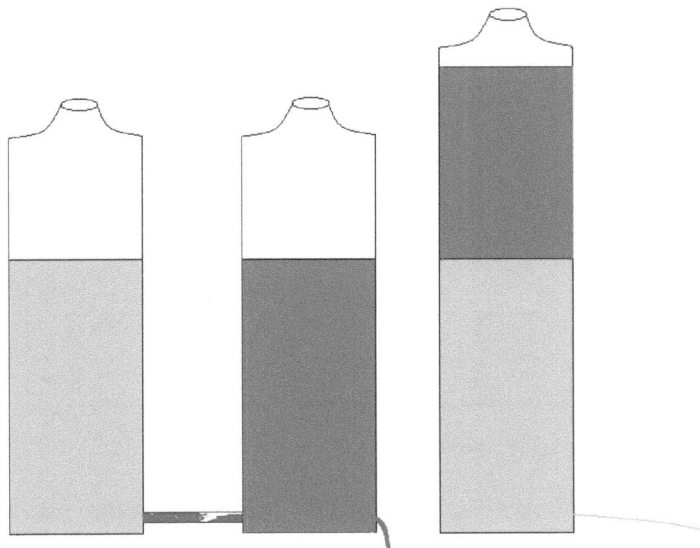

Fig. 2.36 Joined in parallel, same pressure; in series, higher pressure. (Note that in the case of two different heights in parallel, the higher level fluid would flow into the lower level bottle until their pressures equalized)

In the parallel case, the voltage remains the same, at 1.5 V. The reason for this may be found by thinking again of the water analogy. If two tanks are filled to the same height and connected "in parallel," the water pressure (which comes from the height of the water) will be unaltered.

In the series case, we add one voltage to the other—here giving us 3 V. Our analogy is to have one quantity of water directly on top of the other, giving a greater pressure, because of the height, as in Fig. 2.36.

Exercise: A 3 Volt battery is connected in series to a 1.5 Volt battery, and this combination of batteries is connected to a bulb. The bulb has a resistance of 10 Ohms. Calculate the current that flows through the bulb.

(Neglect the small internal resistance of the batteries.)

Answer:
The total voltage (or pressure or height) is (3 Volts + 1.5 Volts) = 4.5 Volts.
Therefore, from Ohm's law, I = V/R = 4.5/10 = 0.45 Amps.

2.26   Elements of Automobile Wiring

Exercise: A car battery (12 Volts in all modern automobiles) has two headlights on, plus six parking lights on—all of course connected in parallel (Fig. 2.37) to the battery.

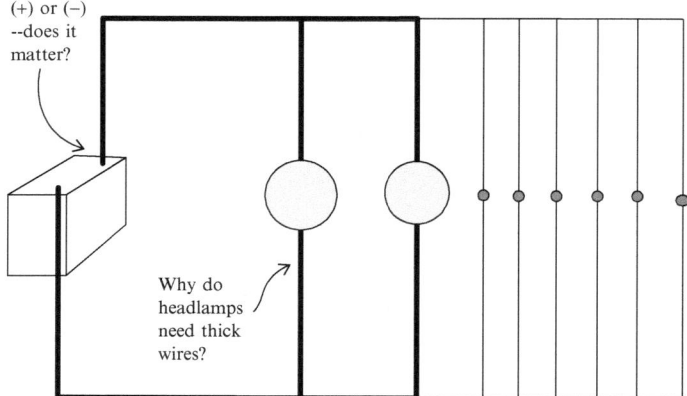

Fig. 2.37 Part of an automobile circuit

The rated power of each headlight on low beam is 60 Watts, say, and each parking light 3 Watts.

If correctly wired in parallel:

1. What is the resistance of one headlamp? (Remember that power = Volts × Amps.)
   Answer: Current through one headlamp is (60 Watts/12 Volts) = 5 Amps,
   and so, from arithmetic,

$$R = Volts/Amps = 12/5 = \underline{2.4\ Ohms.}$$

2. What is the resistance of one parking lamp?
   Answer: Current through parking lamp (3 Watts/12 Volts) = 0.25 Amps, so

$$R = Volts\ /\ Amps = 12/0.25 = \underline{48\ Ohms.}$$

3. Calculate the <u>total</u> current flowing through the battery. (Use Ohm's law, neglecting the internal resistance, or add up the total power.)
   Answer: Total power = 120 Watts + 18 Watts = 138 Watts.
   But 138 Watts = Volts × current = 12 × current, and thus

$$current = 138\ Watts/12\ Volts = \underline{11.5\ Amps.}$$

4. Do you think a 10 Amp fuse (for all these lights on) would be satisfactory? (Such a fuse will burn out at >10 Amps.)
   Answer: No!

5. Also, if the product of current and time ("Ampere-hours") of the battery is given by the manufacturer as 200 Ampere-hours, how long may we expect the battery to last if these lights are left on by mistake?
   Answer: 200 Amp hrs = 11.5 Amps × hrs.
   Therefore, the number of hrs = 200 Amp hrs/11.5 Amps,

$$\underline{\sim17.5\ hrs.}$$

Exercise: Suppose that the two (60 Watts each) headlamps are (incorrectly!) wired in series.

Sketch just the headlamp circuit, and find the total power delivered by the pair. (The power will no longer be the manufacturer's "60 Watts.")

Note that the alternator in a car has to produce more than 12 V—typically nearer 18 V. Also, <u>built-in</u> diodes prevent the charge from leaking back through the windings of the alternator.

All batteries supply current in one direction only. This is called DC or "direct current." There's no such thing as an "AC" battery! Indeed, how would the chemicals be able to "alternate" backwards and forwards?

Exercise: **Multimeter example** (The next 4 pp double as an exercise on series and parallel and may be omitted without loss of continuity.)

While <u>digital</u> multimeters are increasingly popular we look here at the moving coil multimeter or the so-called <u>analog</u> meter. We do this because a digital meter will not help us in the demonstration of, for example, Faraday's law on p. 106 (it would not respond quickly enough) and because the analog meter is an exercise on resistors in series and parallel.

In the current meter we described on p. 43 the force produced by a current flowing in a coil causes a magnet (it was actually a compass needle) to move. In a factory-produced meter it is more usual to fix the magnet and allow the coil of wire to pivot against the restraint of a spring. In fact the mechanism has some resemblance to the electric motor described later (p. 95).

In the photo of Fig. 2.38 the coil lies behind the obscured square at the bottom of the dial, and it has a pointer attached to it which shows the angular movement.

The probes are touched onto points in a circuit wherever current, voltage, or resistance (the relationship, Ohm's law, was given on p. 51) are to be measured, and

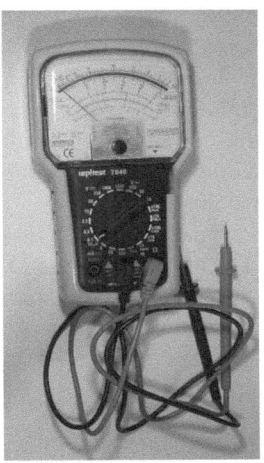

Fig. 2.38 A typical analog multimeter. (Digital multimeters use analog-to-digital converter circuits)

the large black *selector switch* is used to set the quantity you want to measure, as well as the range. (If in doubt, we set it to the largest range!)

The complete instrument depends on the use of resistors—either in series or in parallel. In the circuit diagrams following we have shown only two or three switch positions for each quantity to be measured, but in practice there may be several more.

2.27   Current Measurements

For some of the experiments in this book we will want to measure currents smaller than 100 μA (microAmps)[10] as well as occasionally as strong as 10 Amps.

The moving coil is the heart of our instrument. The strength of the fixed magnet, the number of turns in the coil, and the amount of current we pass through it will dictate how much the coil turns. The limit that it turns through is called the *full-scale deflection* (*FSD*).

Now the coil resistance may be very high, because it consists of many turns and is made from extremely fine copper wire. Many companies make multimeters, and typical values of coil resistance may be anything from about 200 Ohms (a cheap meter) to about 5000 Ohms (a more expensive one.)

For our illustrations let us assume that the coil of the meter has resistance $R_M = 1000$ Ohms.

The only thing that gives us FSD is the current through the coil, and this current is always a fixture for a particular instrument. Let us assume that in our instrument it is, say, **100 microAmps** (or **100 μA, 0.1 mA, or 0.0001 A**).

From Ohm's law $V =$ current $\times$ resistance $= (0.0001A \times 1000$ Ohms$) = 0.1$ V. This means that this particular setting of the selector switch can be for "0.1 V" (and/or "100 μA"). Note that this is a unique case where both a current and a voltage setting are electrically the same. In Fig. 2.39 it corresponds to selector position A. (On a typical multimeter look at the most sensitive current selector switch position; the makers have printed this corresponding tiny voltage reading in parentheses.)

The dashed line represents part of an external circuit in which a current circulates. In order to measure this current you have to break the circuit and feed this current into the multimeter via one probe and release the current back into the circuit via the other probe. We say that the multimeter is in series with the rest of the external circuit.

If you know that it is a tiny current, you move the selector switch to position A which is labeled 100 μA, and then the full current will flow through the meter. (The contact A leads nowhere, but you have to have a mark to turn the switch to!) So if the current happens to be as large as 100 μA, the pointer moves all the way across the scale to FSD; if the current is 25 μA, it moves one-quarter of the way, and

---

[10] Reminder: Milli $= 1/1000$ or a thousandth, and micro $= 1/1000\,000$ or a millionth.

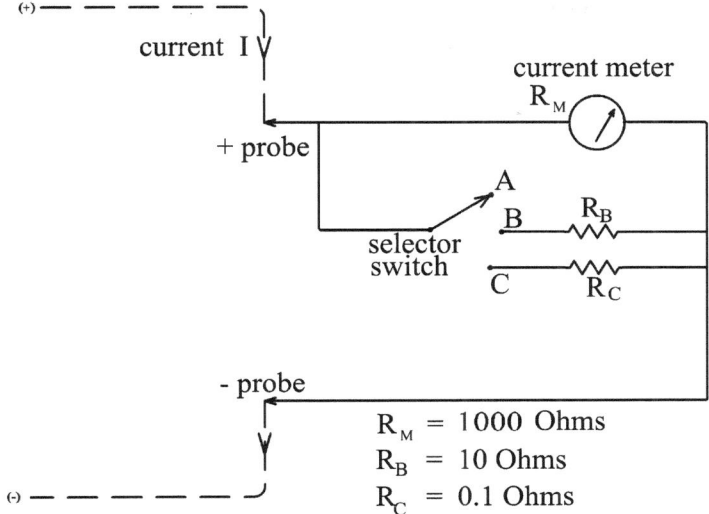

Fig. 2.39 We have cut the wire where we want to measure the current and joined it up again via our meter. For position "A" current flows in at the top left (*red*, (+)) probe, through the meter, and then out at the bottom left (*black*, (−)) probe

so on. Obviously there will also be markings 0–100 on the scale behind the pointer or the needle.

If you want to measure a larger current, without burning out the coil of the meter, it will be necessary to "shunt"[11] much of the current past the coil. Moving the selector switch to, say, position B, some of the current can now also flow through resistor R_B, in parallel with the current through the coil of the meter.

Let's say R_B is 10 Ohms. Then the current through R_B will be a hundred times greater than that through the meter (because we have said that the meter is 1000 Ohms).

If the meter reads the full "100 µA," there will be a current 100 times greater flowing in parallel through R_B, and the total current in the circuit will thus be 100 µA + (100) × (100 µA) or rather 100 µA + 0.01 A. We can ignore the first one (i.e., through just the coil), don't you think? Our meter's coil is now carrying only a hundredth part of the total current—and we can give our scale "B" markings "(0–0.01)A" with negligible approximation.

What happens if we move the selector switch to position C? If R_C has a value of 0.1 Ohms (which is another factor of 100 smaller, *i.e.*, 10 000 times less than the coil resistance) the current through R_C will be 10 000 times greater than that through the meter. Thus we could now measure currents up to 1 A (with yet another scale marked "0–1 A").

---

[11] Any resistor in parallel with a meter is often called a shunt.

Finally, we could measure up to 10 A if we introduced a shunt of 0.01 Ohms—which we might well want to do if investigating an automobile wiring fault for example.

For measuring AC a diode would have to be included.

2.28 Voltage Measurements

Look at the example shown in Fig. 2.40. A lamp bulb is being lit up by the current from a battery. We want to know if the bulb has the correct voltage at its terminals—for if perhaps there is a bad (resistive) connection in the circuit, it won't light properly! So we set the selector to a voltage position (highest volt range to start with, for safety) and connect the probes to the bulb terminals, red probe nearest to the (+) and black nearest the (−).

We know that our multimeter is based on a current meter. Suppose we choose the voltage setting to be D and get a reading on the meter of half of the FSD (which as we know from the previous section corresponds to a current of 50 μA or 0.05 mA going through the coil).

We know that the coil resistance is **1000 Ohms**, and so using Ohm's law we can deduce the voltage between the (+) and (−) probes:

$$V = I \times R, \text{ and so}$$

V = 0.000 05 Amps × 1000 Ohms, which is 0.05 V. (FSD would obviously correspond to 0.1 V, as we saw before.)

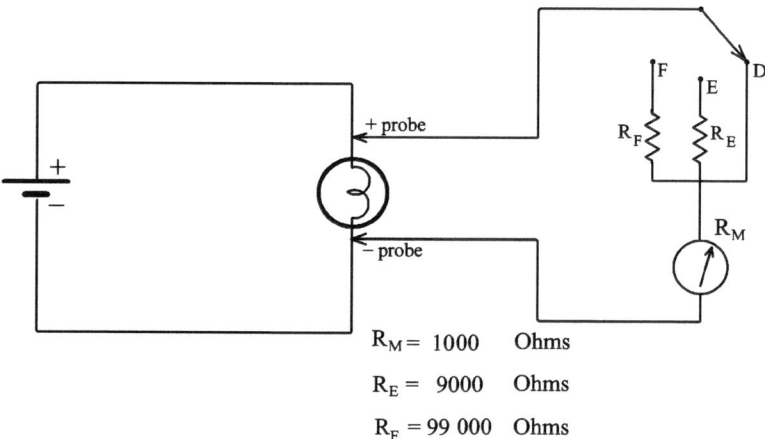

$R_M = 1000$ Ohms

$R_E = 9000$ Ohms

$R_F = 99\,000$ Ohms

Fig. 2.40 A lamp bulb is lit up by a battery. We would like to know the volts across the bulb, so we put our probes there. Note that as always the red (+) probe must be close to the battery (+)

Very often we need to measure greater voltages, but the moving coil of the meter will be destroyed unless we can use just a fraction of the voltage at the probes. We need a potential divider (also illustrated on p. 129).

Let's move the selector to position E. We now have a resistance of 9000 Ohms in series with the 1000 Ohms of the coil. So whatever voltage is at the probes, 90 % of it will be applied to drive current through R_D and the remaining 10 % to drive the same current through R_M. (Equivalently we may say that there is a 90 % voltage drop across R_D and 10 % drop across the coil of the meter.) We have a potential divider of 90:10, and the coil "feels" 10 % of the voltage at the probes. Since our coil can feel a maximum of 0.1 V, the probes can feel up to 1 V. We need a "0–1 Volt" scale, or just read off the "0–0.1 Volt" scale and multiply by 10.

In the same way, with the selector at F, we have a potential divider giving 990:10 or 99:1, and the coil will only feel 1 % of the voltage between the probes. Thus we can now measure up to 10 V.

2.29   Resistance Measurements

We will not be able to measure resistance without an internal battery. This may typically be 1.5 V (or possibly 9 V, or a combination of these, depending on the make).

In Fig. 2.41, R_U is an unknown resistor in part of a working circuit (broken lines) whose resistance we want to measure. Importantly, we must disconnect one end of

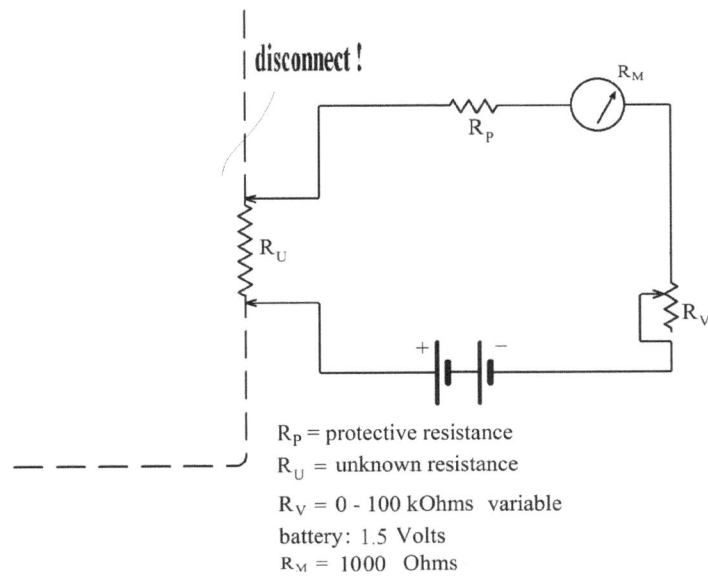

disconnect !

R_P = protective resistance
R_U = unknown resistance
R_V = 0 - 100 kOhms variable
battery: 1.5 Volts
R_M = 1000 Ohms

Fig. 2.41 Measuring an unknown resistance

the resistor. (Without this, there is the real possibility of other currents passing through the unknown resistor, not only distorting our readings but also breaking the meter.)

Note that the internal battery must be able to produce the FSD current of 100 µA. Note also that the resistance scale will run backwards, because a large resistor permits only a small current, and a small resistor permits a large current. We will thus have zero Ohms of resistance at the FSD position and high Ohms at the low end. Further, the markings will not be linear and will appear "squashed" at one end.

For accuracy, because a battery's health slowly dies down, we should set the zero (which is done by holding the two probes together) each time we use the multimeter. Thus, with zero Ohms between the probes, the pointer will zoom over to the right. The variable resistor R_V is adjusted by a small knob or wheel, usually on the side of the multimeter, and you twiddle this until the pointer is at exactly zero on the resistance scale.

Exercise: Figure out, just using Ohm's law as in the two previous examples, an appropriate value for the protective resistor R_P, assuming an internal battery of 1.5 V, probes together, and the "twiddle" resistance R_V adjusted to zero. Also, more ambitiously, think through what the position of the pointer should be if now a resistance of 15 kOhms is the "unknown," R_U.

Answers: [14 kOhms]; [½ FSD].

2.30   Alternating Current and Direct Current (AC and DC)

As we have commented, we know—since the time of the discovery of the electron by the Scottish Nobel Laureate **J. J. Thomson** (1856–1940) in his cathode ray experiments of the 1890s—that it is the *electrons* that move down a wire. However, as we mentioned earlier it doesn't really matter whether we assume the *mathematical* current from (+) to (−) or the *physical* current from (−) to (+).

In all households, the current supplied by the utility company flows back and forth, changing direction very fast. One moment the voltage polarity on the two parallel "slots" of the wall outlet is "+ −," and the next moment it is "− +," as in the sketch on the right-hand side of Fig. 2.42.

The right-hand diagram of Fig. 2.42 shows one *complete cycle*. In the first half of that typical cycle the voltage is positive, and in the second half it is negative. This means that current through a device will flow backwards in the second half and will go on alternating in direction until switched off.

How do we create AC? It comes from the steady rotation of a coil between the poles of a magnet. This is an example of **electromagnetic induction** which we discuss in detail on p. 106. It was the Serbian, **Nicola Tesla** (1856–1943), once employed by Edison, who was the major pioneer of this type of current. Indeed, when Niagara Falls was harnessed, it was Tesla's beautifully designed AC generators that were installed there by the Westinghouse Company in the 1890s.

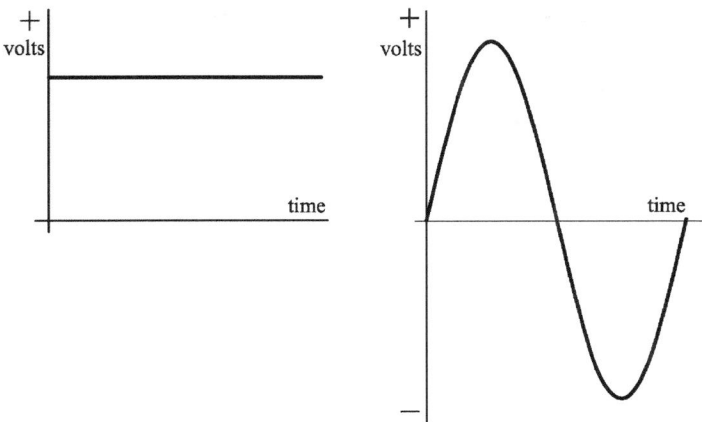

Fig. 2.42 DC voltage; AC voltage

Edison was a DC enthusiast, but DC lighting systems have severe disadvantages—notably resistance losses—energy wasted as heat (Joule heating) when the current flows through cables over long distances. Because AC can be "transformed" (see "the transformer" in Chap. 3, p. 109) into high voltages and <u>low currents,</u> reducing resistive losses, it didn't take long for AC to win out.

Power companies in the USA supply power at 60 complete waves, oscillations, or cycles per second, but in many other parts of the world power is supplied at 50 cycles per second. This is called the *frequency* and is measured in "cycles per second" or "Hertz," abbreviated to Hz, in honor of **Heinrich Hertz** (1857–1894). In 1887 Hertz used sparks to produce Maxwell's theoretically predicted radio waves. He died far too young, at the age of 36.

As mentioned, the coils of a generator rotate only 50 times a second in the UK, for example. This means that some American devices will not work properly in many other parts of the world. A clock designed in America to "tick" 60 times per second will only "tick" 50 times in the UK, thus running consistently slowly. And there are other differences: European voltages are all ~240 V, whereas most American devices like shavers and so on run from 120 Volts. This means that, connecting an American shaver, for example, directly (without a transformer) to a European outlet will burn it out.

You may be wondering what the value of the voltage "really" is, since we see from the previous graph that it varies up and down between + and − 170 V (in the USA) and + and − 340 V in the UK and many European countries.

So we need some kind of average, but wouldn't the average of these up-and-down variations be zero? After all, if one pushes a child back and forth on a swing, the "average" position is neither in front nor behind—it is at the center or at zero displacement.

So what we do is define an average based on the amount of <u>energy</u> produced, for example, in a light bulb—after all, energy is being converted whether the current is

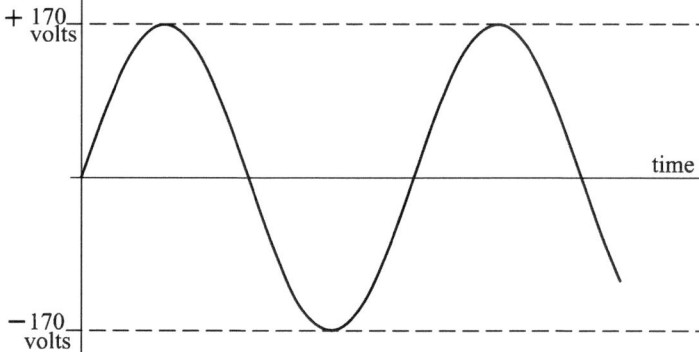

Fig. 2.43 The actual AC kitchen wall outlet varies between +170 and −170, yet we call it a 120 V supply. The reasoning behind this is indicated briefly in Appendix F

flowing backwards or forwards. After such calculations are done, the average value for voltage that people use turns out to be about 2/3 of the maximum voltage.

More exactly, we actually divide the maximum by the *square root of 2* (*i.e.*, 1.414), so our average value of voltage (again, in the USA) is[12] (170 V)/1.414 = 120 V.

The wall outlet voltage is sketched in Fig. 2.43. If we could examine the voltage in slow motion and connect a meter [**do NOT attempt this!!—the voltage is dangerously high**] we would in principle (but not in reality—it's too fast) see the needle of a center-reading voltmeter oscillating back and forth. Fortunately, AC scales on voltmeters read these average values, so we can completely forget about maximum values in everyday life.

2.31 Skin Effect

In a wire carrying AC nearly all the current is concentrated at the outside, and the higher the frequency of the oscillations the more pronounced is this effect. People refer to it as the *skin effect*. It does NOT take place with DC.

Clearly, in the kitchen, we don't care whether the current in our wires is in the center or near the outside, but our AC cables could in principle be hollow. However, this is not particularly meaningful at the rather low frequency of the mains—60 or 50 Hz, depending on the country—and the skin depth at these frequencies is more than a centimeter (or the thickness of a finger). Most household cables are less than the size of a finger, so any power losses are therefore negligible.

Photographs of high-power transmission lines occasionally show double cables, because engineers know that since the alternating current is restricted to the outsides,

---

[12] Any reader who may be familiar with calculus and trigonometry can see a reminder of the basic reason for the square root of 2, given in small print in Appendix F.

it makes sense to sometimes put in two wires, rather than one thick one. Thus we are utilizing two skins rather than the skin of one heavy and thicker cable.

This is also the reason (or one of the reasons) why some wires are stranded—the so-called *Litz* wire. Again, we then have lots of little skins rather than the single skin of a heavy wire.

Yet another reason is that stranded wire is more flexible.

2.32 An AC Experiment with LEDs

An LED will also run off AC because it is a diode. It will light up on alternate half cycles, as in Figs. 2.44 and 2.45.

If we were to connect 50 or 60 little LEDs in series they would light up directly from a regular 120 V AC outlet.

They are lighting up much too fast, of course, for us to see them switching on and off.

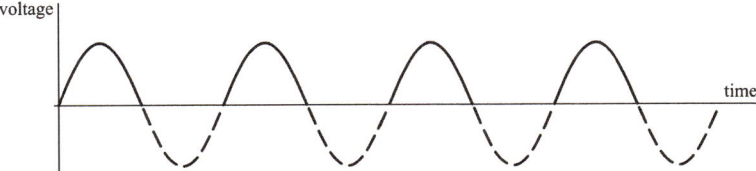

Fig. 2.44 On AC power, an LED will switch on and off every cycle—too fast for the eye not to believe this to be continuous

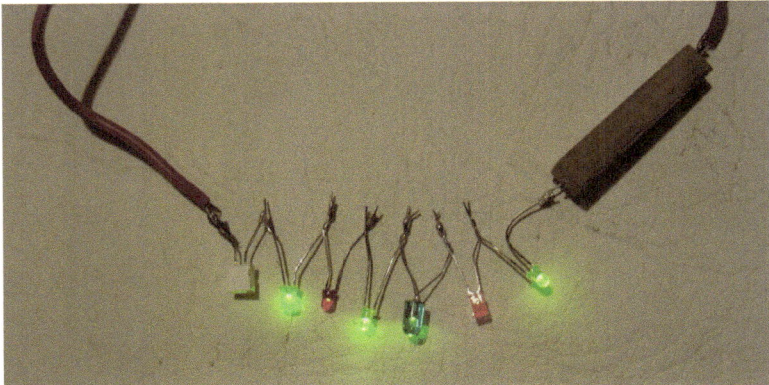

Fig. 2.45 Seven small assorted LEDs soldered together, in series with 5600 Ohms, running directly off the regular 120 Volts AC. The current flowing, from Ohm's law, would then be approximately 22 mA or 0.022 Amps. Note that a couple of the randomly selected LEDs are quite faint

Also, should one of them burn out they will all turn off, because the current would be interrupted. (Some old <u>incandescent</u> Christmas tree lights may not do this because they may have been uniquely designed to "melt" and yet leave some continuity.)

In Fig. 2.45, instead of 60 of them, 6 or 7 have been soldered together in series with a large resistance of 5600 Ohms.

CAUTION: The home experimenter should <u>not</u> try this! 120 Volts or more is too dangerous to even think of playing with without very great care.

Magnetism

3.1   Lodestones

Historically, the iron ore **magnetite**, or **lodestone**, which is found in various parts of the world, especially in Magnesia, Greece, was known over 2000 years ago to swing around if suspended on a string.

By the thirteenth century it was realized that a sphere of lodestone floating in water (perhaps on a piece of wood) acted as a compass. **Peter de Maricourt** (exact dates unknown), a French scholar and experimenter, also called "Peter the Pilgrim," wrote to a friend:

> "... you'd be able to direct your steps to cities and islands, and to any place whatever in the world ..."

He also noticed, around 1270, that a steel needle placed anywhere on the surface of a lodestone would align itself in a unique direction, pointing to a "pole" at the top of the lodestone or from another "pole" at the bottom. He called the ends of some of his needles "north-seeking" (N) and "south-seeking" (S).

By making quite a few "lodestone spheres" Peter the Pilgrim noticed that *like poles repel* and *unlike poles attract*. Now just as a charged comb can induce charges on scraps of paper and attract them, as we saw in Fig. 1.12, so a chunk of lodestone, or merely a kitchen magnet, can magnetize certain materials nearby, as in Figs. 3.1 and 3.2.

We mentioned Dr Gilbert, physician to the Queen of England, at the beginning of the book (p. xi). Around the year 1600, shortly before he died, he suggested that the earth itself was like a large spherical lodestone, and he drew sketches of what he described as *"lines of magnetic virtue"* around a magnet. If you sprinkle iron filings

---

The online version of this article (doi:10.1007/978-3-319-05305-9_3) contains supplementary material. This video is also available to watch on http://www.springerimages.com/videos/978-3-319-05304-2. Please search for the video by the article title. The supplementary audio material can also be downloaded from http://extras.springer.com.

D. Nightingale and C. Spencer, *A Kitchen Course in Electricity and Magnetism*,
DOI 10.1007/978-3-319-05305-9_3, © Springer International Publishing Switzerland 2015

Fig. 3.1 Three screws
(initially unmagnetized) hang
from a regular refrigerator
magnet

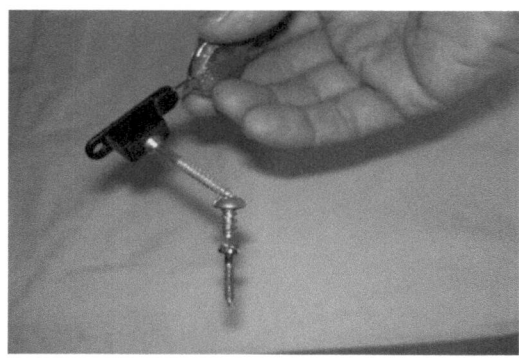

Fig. 3.2 The permanent
kitchen magnet picking up
screws. We don't yet know its
"polarity," *i.e.*, which side
is N or S, but have labeled
things arbitrarily for
simplicity. We find that the
screws behave in a similar
way to the scraps of paper
in Fig. 1.12

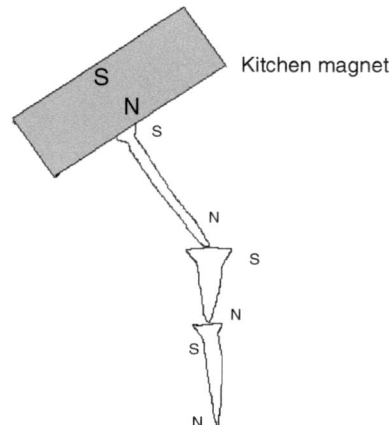

(just take a nail and file it down) onto a piece of paper which is covering a magnet you will see his lines of virtue. Such lines of magnetic virtue represent the directions of what we now call the magnetic *field* or what physicists call **B**. They show the directions that a free N pole, if there were such a thing, would move along—analogous to the **E** field discussed on p. 18.

Note that the **B** field lines are <u>continuous,</u> looping round for ever and passing through the interior of the magnet. This is different from the electric dipole. <u>If we had drawn the electric field (</u>**E**<u>) for a dipole (which we leave as an exercise) we would have found the same shaped external field as the above. However, E fields begin and end on (+) and (−) charges, whereas B field lines are seen to be continuous loops, running through the magnet itself—they have no beginning or end.

This fundamental difference between magnetism and electricity implies that there is no such thing as an isolated North or South pole. In Fig. 3.3 it *looks as if* the magnet has poles, but if you were to cut one of the poles off you would not find an isolated magnetic pole or a "monopole," but rather two magnets and thus four poles, as in Fig. 3.4.

Fig. 3.3 Sprinkling iron
filings on a piece of paper (the
magnet would be under the
paper) or plotting around with
a little compass will show the
directions of the magnetic
field **B**, which by convention
is away from the N and
towards the S. (This is
analogous to the convention
in electrostatics, where the
direction of **E** is away from
the (+) and towards the (−).)
Also, where lines are bunched
the field is strong

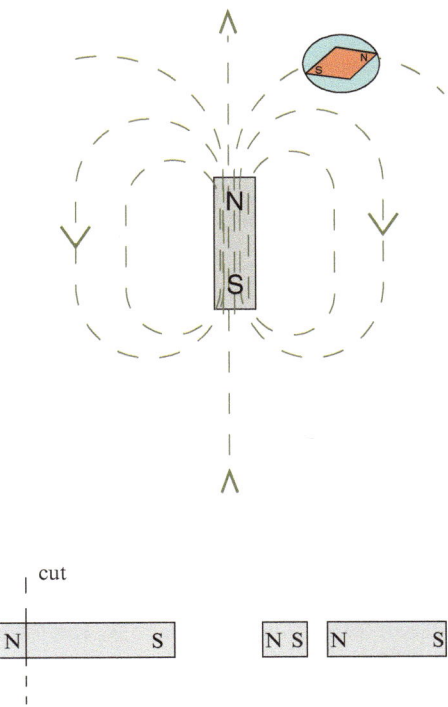

Fig. 3.4 Failing to isolate
a single "magnetic pole."
All we get is another magnet

This mystified people for a long time. Worse, in the twentieth century a highly respected theoretical physicist, **Paul Dirac** (1902–1984), a 1933 Nobel Prize winner, suggested, for rather erudite reasons, that IF there were a small number of monopoles in the universe, this would be of theoretical interest. We will not go into this, but we note that monopoles have never been observed.

Meanwhile, the well-established equations of classical electricity and magnetism, especially the four equations given by the famous Scottish physicist **James Clerk Maxwell**, given in the Appendix (p. 170), have never been found to be wrong. We will definitely not need his equations, but our book does illustrate their significance in various places, with the appropriate reference pages given in that Appendix.

The subject of electricity in general owes a great deal to Maxwell. His family commented about their 3-year-old boy in a letter this way: "he is very happy, and has great work with doors, locks, keys etc and '*show me how it doos*' is never out of his mouth." At school he was shy and made no friends, but at about 14 he suddenly began gaining prizes for mathematics as well as for verse. He graduated from Cambridge at 23, and although he never reached 50 his admirable mind enriched many branches of physics, including optics and thermodynamics.

Our impractical kitchen gedanken idea in Fig. 3.4 (*gedanken* is the German word we met in Chap. 2 to describe a thought experiment) illustrates that a magnetic pole cannot be isolated. However hard we try to cut a pole off, we won't succeed!

Fig. 3.5 Our earth behaves like a big bar magnet. The magnetic poles are not exactly at our geographical poles. The imaginary magnet is shown in *grey*, and note that its S is at the <u>north</u> end. Why? Because our compasses line up with their N pointing north, and they can only do that if the imaginary bar magnet is as shown

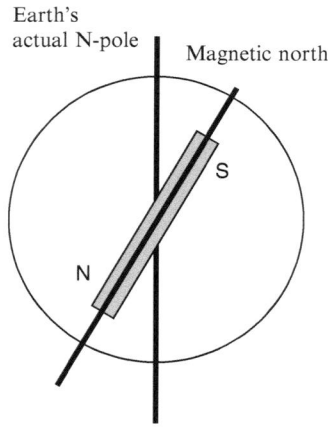

Earth's actual N-pole

Magnetic north

3.1.1 The North

The earth's magnetic field is not aligned precisely with the earth's unshakeable axis of spin, and it is the latter that defines earth's actual poles. The difference is shown (exaggerated) in Fig. 3.5. The location of the *magnetic* north according to a regular household compass varies slowly over time and is currently near Canada's Hudson Bay.

Similarly, our ordinary compass will point to the south as being somewhere in Antarctica, a little south of Tasmania.

It is thought by many researchers that the earth's field has flipped, north for south, possibly hundreds of times—perhaps as recently as 30 000 years ago. Furthermore, since the interior of the earth is too hot for iron and other magnetic materials to remain magnetic—heat destroys magnetism—it's clear that we are dealing with something other than a bar magnet inside the earth!

One possible explanation (among many) for the earth's magnetism is that the molten iron is ionized, and the moving ions cause the **B** field.

3.2 Further View of Magnetism

It is known that magnetic materials have regions called *domains*, actually visible by means of electron microscopy. Unmagnetized, the domains of a material are as on the right of Fig. 3.6, and when magnetized the domains acquire a "commonality of magnetism" as shown on the left of Fig. 3.6.

To *de*magnetize something either it can be heated to a high temperature (above the so-called Curie temperature, ~1400° F for iron—much hotter than in the kitchen oven) or it may be tapped repeatedly with a hammer. The thermal agitation, and/or

Fig. 3.6 Magnetic domains.
A magnetized sample on the
left, and not magnetized on
the *right*

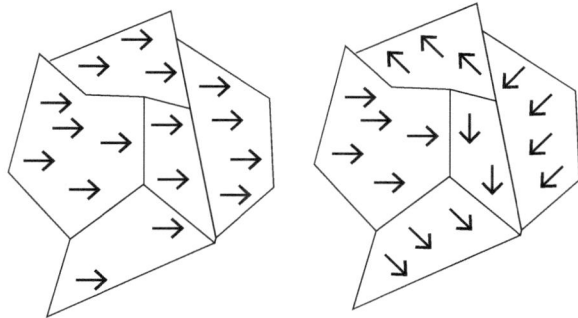

the tapping, tends to return these domains to more random orientations, again as on
the right of Fig. 3.6.

3.3   A Kitchen Compass

We can easily make a compass.

Take an old hacksaw blade, <u>wrap it in cloth</u> to avoid any possible shards, and
snap it in half (hacksaw blades are brittle enough for this).

Take a refrigerator magnet (or any magnet, the stronger the better), and stroke it
along one of the halves of the hacksaw blade about a dozen times, in one direction,
as shown in Fig. 3.7.

Then suspend the hacksaw blade on a piece of cotton or string (Fig. 3.8). The
blade will swing around and slowly settle in a north–south direction.

Our compass took less than 5 min to make. It is most certainly pointing north, as
checked by a regular compass.

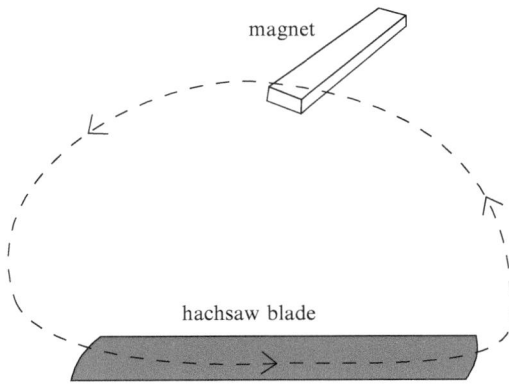

magnet

hachsaw blade

Fig. 3.7 Stroking the
hacksaw blade <u>one way</u>

Fig. 3.8 Homemade
compass

3.4   Angle of Dip

If we look back at Fig. 3.3, the field lines "dip" towards the magnetic poles, and so if
the earth behaves like a magnet as in Fig. 3.5, then a compass needle (or hacksaw
blade) will "dip" down towards the current region of the earth's poles.

Notice that our hacksaw blade in Fig. 3.8 is not far enough north to be dipping
noticeably, and, to get an accurate measurement, the hacksaw blade should be
suspended so that there is no hindrance to any dipping motion—which is not quite
the case in our photograph.

The dip angle is the angle between the axis of the magnet and the horizontal, as
shown in Fig. 3.9.

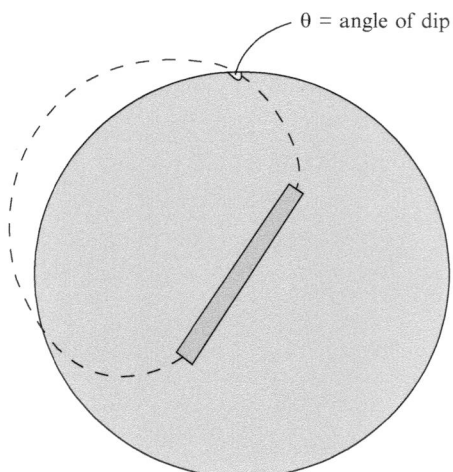

θ = angle of dip

Fig. 3.9 The angle of dip, θ

3.5 Diamagnetism

Many substances are slightly *repelled* by magnets and are called *diamagnetic* [Latin, *dia*, "against"]—a name coined by Michael Faraday.

Water, glass, rubber, copper, wood, and many other substances are diamagnetic, but the effect can only be seen with <u>extremely</u> strong magnetic fields. Such fields are essentially unobtainable in the kitchen, so the minuscule repulsion that should be observed in Fig. 3.10 is not possible to detect at home. It leaves one wondering how Faraday detected it.

However, we would have a bit more success if we used pyrolytic graphite. Inexpensive kits are sold for this kind of demonstration on the Internet—kits which typically include a thin sheet of graphite shown levitating about a millimeter above an array of half a dozen powerful neodymium magnets.

Fig. 3.10 Water, which is diamagnetic, *just* being repelled by a magnet. Unfortunately, the effect is not measurable without an extremely strong magnet, preferably in a laboratory setup

3.6 Paramagnetism

Substances that are *slightly* attracted by B fields are called *paramagnetic*—for example, platinum, sodium, aluminum, and some gases. All these substances have an odd number of electrons, so that each atom, as we will see shortly, is able to act like a little magnet. Such materials are drawn towards a powerful magnet, but only very weakly.

3.7 Ferromagnetism

The substances we discussed earlier—notably iron and nickel and steel, like our hacksaw blade—show strong magnetic effects and are called *ferromagnetic* [Latin, *ferrum*, iron]. In these substances the little atomic magnets in their domains <u>and</u> the domains themselves *align cooperatively* and so are strongly attracted to an external magnet.

Magnetic recording tape and computer hard drives consist of a thin layer of such material, usually on a plastic substrate.

Sometimes it is possible to detect fake coins in the kitchen. If you slightly tilt a large coin and let a very small neodymium (see p. 119) magnet—smaller than the coin—slide down it, the following can happen:

1. Coins with any iron in them (most likely cheap fakes) will cause the tiny magnet to slide down sluggishly or stop.
2. The magnet may slide down unaffected, not interacting appreciably. For example, these may be copper coins. However, as with Arago's disk (p. 119), there should be some <u>small</u> eddy currents, which could slow the slide slightly.
3. Coins with gold and/or silver will get slightly more eddy currents in them and thus should slide down a little slower still.

(Note that many so-called copper and nickel coins are in fact iron or steel or zinc with a copper or a nickel plating: for example, in the USA, since 1982; the UK since 1992; the Euro since inception; and the Canadian nickel.)

3.8   Shielding

A magnetic field can be shielded by certain materials, and this we can show in the kitchen.

Iron and steel are very *permeable* to magnetic fields, which means that they are very *accepting of the lines of force*. Magnetic fields tend to concentrate in these materials, and so we say that the materials have high *permeability*.

If we surround a magnetic compass with iron, as in the right-hand side of Fig. 3.11, the field should be concentrated in the iron itself, leaving a negligible field beyond.

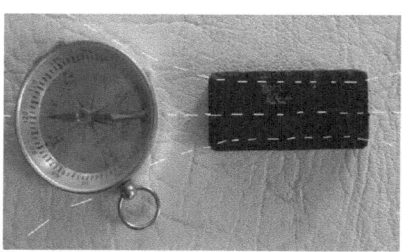

Fig. 3.11 The compass (*left photo*) is affected strongly by the field of the kitchen magnet, but in the *right photo*, an iron cooking pot has shielded the compass, which now points in no particular direction

Typical values of permeability, taking the permeability of air as our reference, are:

[Air	= 1]
Aluminum, copper, rubber, paper	~1
Steel	~1000
Wrought iron	~2000
Permalloy (78 % nickel, 22 % iron)	~80 000

To visualize: What this means is that with, for example, permalloy, 80 000 field lines (or B lines) will be concentrated in the region where there was only one line before.

Now try placing a kitchen magnet a few inches from a compass, and see how it completely swamps the magnetic field of the earth. The left-hand side of Fig. 3.11 shows this.

Then take a large iron (aluminum and/or copper are no good—see the table above) cooking pot, and let a compass swing inside it, as in Fig. 3.11 (right photo).

The compass needle no longer aligns itself with any particular direction, because the iron pot is a shield. The magnetic field lines, *i.e.*, the **B**—or what Dr Gilbert would have called the lines of virtue—are concentrated in the (high permeability) cooking pot. Note that the metal frame of the compass is most likely made of aluminum, although it looks deceptively like steel (again see the table above).

3.9   Different Magnet Shapes

Magnets are made in many different shapes, such as those sketched below, and they can even be flexible (Fig. 3.12).

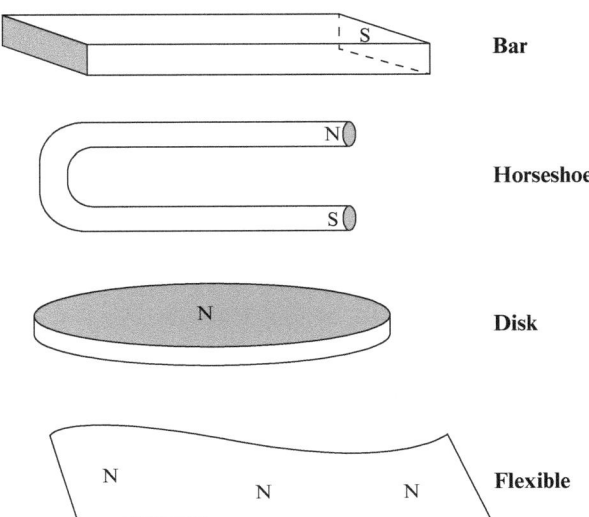

Fig. 3.12 Some magnet shapes. Manufacturers may choose the directions of the fields

Common refrigerator ceramic magnets are made by heating powdered iron oxide with, typically, barium carbonate ceramic (to above the Curie point—see p. 80), putting the hot mix in a strong field (in any direction) and letting it all cool. The domains will then stay aligned, and we have a permanent magnet.

Alnico magnets (a mix of aluminum–nickel–cobalt) possess somewhat stronger fields, and magnets that contain neodymium (mentioned further on p. 119) are stronger still.

One may visualize the shape of the **B** field around any magnet by considering how a little test compass, as we saw in Fig. 3.3 (p. 79), would behave.

3.9.1 Aurora Borealis

While this certainly is not a "kitchen experiment," the aurora (*Aurora* was the goddess of dawn) is a pretty illustration of the presence of the earth's magnetic field. *Northern lights*, commonly called the *aurorae borealis* (*borealis* means "of the North") are due to electric charges (*e.g.*, protons, electrons) from the sun, trapped and circulating in the magnetic lines of force of the earth. (We introduce the Lorentz force on p. 92.) We can't see the charges coming from the sun, but when they collide with air molecules, light is given off, of varying colors and great beauty.

3.9.2 Magnetic Bacteria

It was noticed, in 1963, by a microbiologist in Italy at the University of Pavia, **Salvatore Bellini**, and independently in 1975 by a Woods Hole microbiologist **Richard Blakemore** that some bacteria act exactly like little magnets. When a magnet was brought near to them they were seen to drift.

They can follow the earth's magnetic field lines, and this has a use. We have mentioned the angle of dip—and since bacteria don't like oxygen, they can follow the earth's field lines away from the equator down to deeper parts of the oceans where there is less oxygen. As magnets they can't avoid following the field lines!

3.9.3 Tapes and Swipe Cards

A cassette (audio or video) has tiny magnetized iron particles on the moving tape, and thus there can be (either analog or digital) information stored on the tape. Also, computer hard drives, as we mentioned, consist of a hard platter of glass or aluminum, with a similar type of magnetic coating. Changes in the magnetization in these films can represent music, voice, or other data.

A swipe card (*e.g.*, a credit card or a library card) has magnetized particles on the back which can contain data on name, address, phone numbers, etc. (However, it should be noted that CDs and DVDs use optical methods—nothing magnetic about them at all.)

Obviously, a strong magnet would be able to demagnetize a tape. One could take an old unused tape and try this. However, it will take a much stronger magnet than a fridge magnet to achieve complete demagnetization.

3.10 What Causes a Magnetic Field?

In 1820, the Danish scientist **Hans Christian Oersted** (1777–1851)—who befriended the slightly younger writer **Hans Christian Andersen** (1805–1875)—discovered something that turned out to be quite profound. While experimenting, he observed that an electric current affected a compass (Fig. 3.13). Although nature seems to provide us with magnetism with apparently no associated currents, as in lodestone, it is known now that such an idea is not true: all magnetism has to come from currents.

Such magnetism is what we found when we made our "amp-meter": the current through our coil caused the compass needle to come to a new position as in Figs. 2.14 and 2.15.

Let's repeat Oersted's fundamental history-making experiment directly—in the kitchen of course!

3.11 Oersted's Experiment

Place a wire above an ordinary household compass, as in Fig. 3.13, and touch the ends briefly to a small battery—*e.g.*, a 1.5 V battery. This is a direct short circuit, so it must be brief! If you leave it on for more than a second the wire will get hot to touch and the battery will very soon be completely depleted.

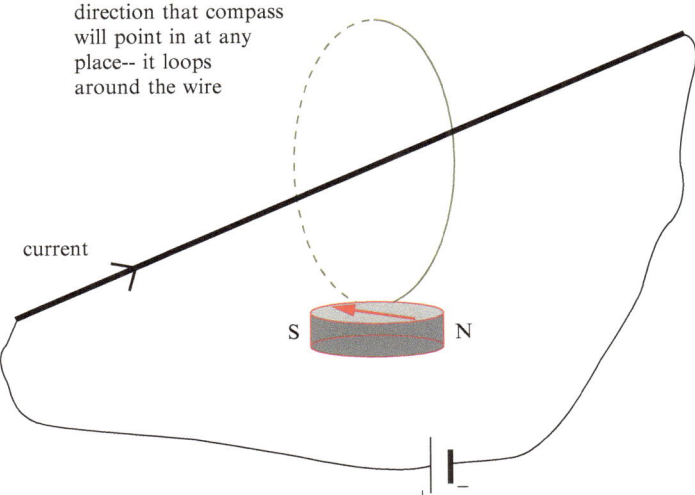

direction that compass will point in at any place-- it loops around the wire

current

S   N

Fig. 3.13 Oersted's experiment. The compass originally pointed north. Now, with the strong current, the needle points in a new direction, which is always around the wire

(This is a case where the fat 6 V battery, mentioned on p. 157, would be useful.)

Note the directions, for the compass needle must of course swing so as always to align itself along the *new magnetic field*, **B**.

By investigating a few different positions of the little compass (again tapping the short circuit as briefly as possible) we soon see that the directions of the field lines are as shown by the brown circle around the wire of Fig. 3.13.

Moral: The current obviously does produce a magnetic field, and it goes around the wire. Also while we don't need this here (it may be found in the Appendix) it was the great **Andre Ampere** (p. 29) who described Oersted's observation with an exact mathematical formula.

Further experimentation shows that currents and their magnetic fields are inseparable—a fact we mentioned in the Preface. Of course with tiny currents you only get tiny fields—*e.g.*, an electron orbiting in an atom—but they have to be there. Oersted's demonstration has just said so!

The maximum effect on the compass needle will be observed when the wire is originally set in an N–S direction, *i.e.*, parallel to the compass needle.

Now in Fig. 3.13 imagine your thumb (using your right hand only!) pointing along the wire <u>with</u> the current.[1]

The directions of the magnetic field lines are then those of the fingers of the right hand; that is, the field lines curl around the wire as the fingers would. (Note that in some books the left hand is used—but this is only if the current is considered to be *electron* current. It is only necessary to use the right hand in physics.)

3.11.1 Shape of the Field Due to a Loop

If we bend the wire of Fig. 3.13 into a big single loop, we can fairly easily see that the **B** field takes on the shape shown on the left-hand side of Fig. 3.14. For many loops the field lines, shown on the right-hand sketch of Fig. 3.14, will be concentrated mostly in the center of the coil.

3.12 A Coil

Thus, the more loops we have, then Oersted's effect will increase, as has just been shown in the right-hand sketch of Fig. 3.14.

Take a straw (Fig. 3.15), cut it to about 4″ in length, and wind insulated wire (maybe 20 turns) around it. This coil is commonly called a *solenoid*.

---

[1] Not the electron flow, but the mathematical current from + to −.

Fig. 3.14 Bending Oersted's straight wire into one big loop, the fingers will now curl through the center, as the field lines show, and many loops will exaggerate this, resulting in a relatively uniform field through the center of the coil

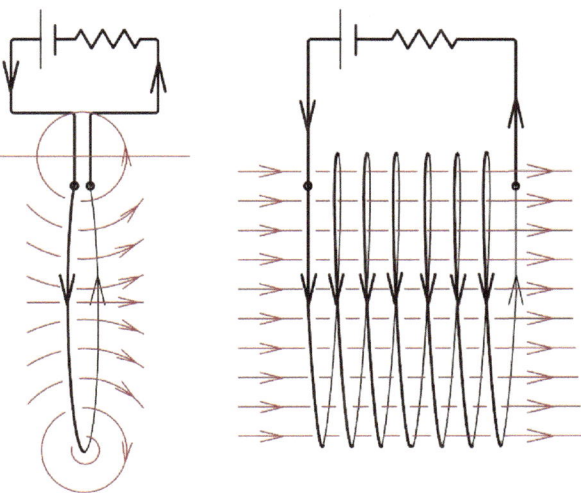

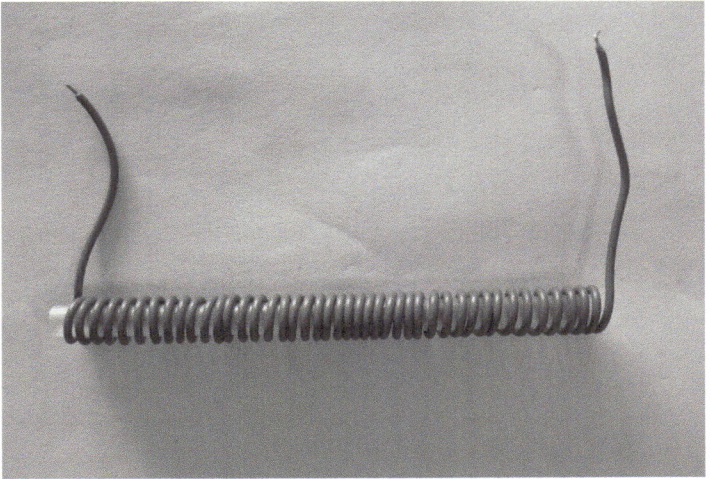

Fig. 3.15 A coil of wire or a solenoid

If we now temporarily short-circuit an old battery across the solenoid's ends (again be very brief because things will otherwise get way too hot!) a compass at either end will be affected. In fact, the magnetic field produced by this solenoid is of the same configuration as that of the field we found for a magnet in Fig. 3.3.

Alternatively, put in a resistance (such as a bulb) to limit the current, as has been indicated in Fig. 3.14, and then there's plenty of time to observe the compass in

Fig. 3.16 The iron or the steel core on the *right* yields a much stronger magnetic field. The domains in the iron are now lining up, increasing the field from the solenoid. (Same current on the *left* as on the *right*)

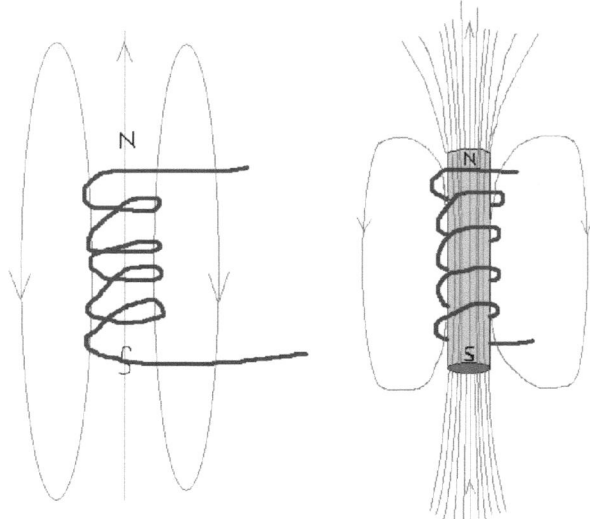

various locations without the wire getting hot. The disadvantage of course is that our probing compass will then experience smaller deflections.

3.12.1 Experiment

Now place a long nail, or a thin screwdriver or a thin iron rod, inside the solenoid. The magnetic field (B) will become stronger (more concentrated)—as mentioned before and as shown on the right side of Fig. 3.16. The effect on the nearby compass will be more pronounced.

This is the basis of the electromagnet. **Joseph Henry** (p. xi), working in Albany, NY, made one of the first really strong ones, capable of lifting almost a ton without a lot of current. The extra strong field comes from the lining up of the domains shown in Fig. 3.6.

Many things use solenoids—and one of the most common is a *relay*.

Relays are used to switch on other, more "powerful" circuits which are to carry larger currents. In an automobile for example, when the ignition key is turned, it sends a small current to a "starter relay," which then allows a huge current to go through the starter motor. If such a huge current went through the ignition key circuit, the wires would melt.

A relay is sketched in Fig. 3.17.

In the case of the automobile starter, "s" connects to "B" and would then allow a large current to flow along a thick wire to the starter. ("A" here would play no part, but could play a part in other situations.)

Fig. 3.17 The spring keeps
"*s*" touching "*A*." However,
when a small current is sent
through the solenoid, "*s*"
(made of steel) will be
attracted to the solenoid
and will thus touch "*B*"

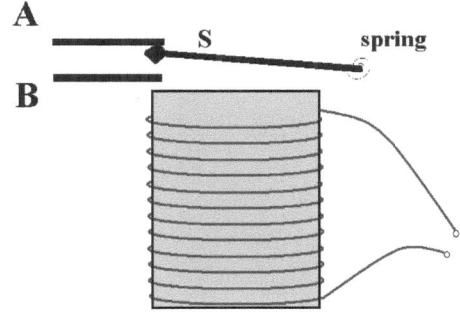

3.13   Inductance (L)

There are about 40 turns of wire on the straw of Fig. 3.15. A DC voltage and
consequent current will cause the creation of a magnetic field B, and that current
will be very large because the copper wire offers no appreciable resistance.

Now nineteenth-century experimenters found that a varying voltage across the
coil (*i.e.*, AC instead of DC, as on p. 73) caused less current than expected and
surmised that a strange induced voltage, in opposition to the voltage we applied to
cause the current in the first place, must have formed. This is sometimes called a
back emf.

The very fact that the coiled wire is producing a varying **B** seems to be the cause
of this *induced* opposition to expected current flow. We say that the coil has
inductance, L.

Inductance L is measured in Henries, in honor of Joseph Henry, and for any
specific coil it depends on size and shape. The L of the solenoid-shaped coil, such as
in Fig. 3.15, turns out to depend very strongly on the number of turns and quite
strongly on the diameter of the straw. Thus for the 40-turn coil of Fig. 3.15, a fatter
coil would have a bigger value of L than a thin one of the same length.

The actual formula for the inductance of a solenoid is given in the Appendix.

3.14   A House Alarm

A similar thing to a relay is a "magnet switch," sometimes called a "reed switch."
This is the small object on the left of Fig. 3.18.

The switch "s" of the above relay could just as well be operated by a little fixed
magnet instead of the above relay coil.

If such a little magnet were embedded in a window, or a door, for example, it
would be easy to "make" or "break" a circuit, and a bell or an alarm could be made
to ring.

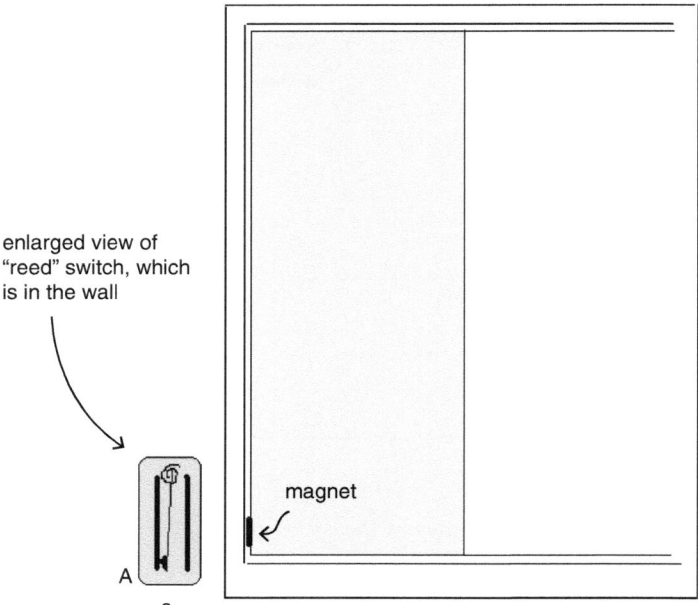

Fig. 3.18 A small magnet is fixed in the sliding door. If the magnet is close enough to the "reed switch" it might break the connection of "*s*" to "*A*." (In the sketch it is not quite close enough yet!)

We have seen that a current always causes a magnetic field. What happens if we have another magnetic field near to a current? We will find that there is a force—and either the magnet or the wire will move.

3.15 Experiment: Force on a Current Near a Magnet (Lorentz Force)

In the photo of Fig. 3.19 we have a 3 V bulb (to be driven by two 1.5 V C or AA cells) in series with a <u>lightweight</u> enamel-covered wire hanging in front of and not quite touching a strong kitchen magnet. This particular magnet has its poles on the flat faces, and so the field is directly out towards us, if we are standing in front of the refrigerator door. A tiny chalk mark on the magnet will help show the movement of the wire.

When the switch (on the floor) is turned on, the little bulb will light and the vertical wire will be kicked sideways.

It will stay away, showing that there is a force all the time the current flows.

Fig. 3.19 A wire hangs
vertically down from a
flashlight bulb, just in front
of a strong kitchen magnet
on a refrigerator door.
There is a little *chalk mark*
on the magnet

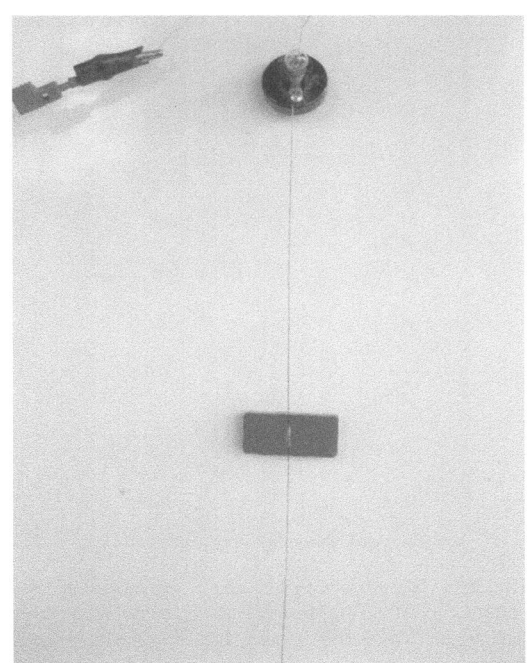

3.16   Direction of Lorentz Force

Figures 3.19 and 3.20 show what happens, and there's an easy way to summarize
all this. It's again a *right-hand rule* (Fig. 3.21), and it shows the three mutually
perpendicular directions involved.

We might have already observed that the current-carrying wire and the B field
from the permanent magnet in Fig. 3.19 (or Fig. 3.20) had to be mutually perpendicular for the best effect. (Further experimentation would show that if they were
parallel there would be no effect.) Recall that the field here was coming out from
the refrigerator towards us HORIZONTALLY, and the current in the wire was
VERTICAL. Figure 3.21 gives the direction of the force. (Thu**M**b [**M**athematical
current], **F**irst finger [**F**ield]; the center finger gives the direction of the consequent
force.)

This force, called the Lorentz force after the Dutch physicist **Hendrik
A. Lorentz** (1853–1928), is why any electric motor works. His law states that
there is a force on any current-carrying wire that is near a magnet.

The young Lorentz was quite precocious. Son of a nurseryman, he entered the
University of Leyden at 17, obtaining his bachelors degree in mathematics and
physics at 18! Teaching night school he obtained his Ph.D. (in optics) at 22.

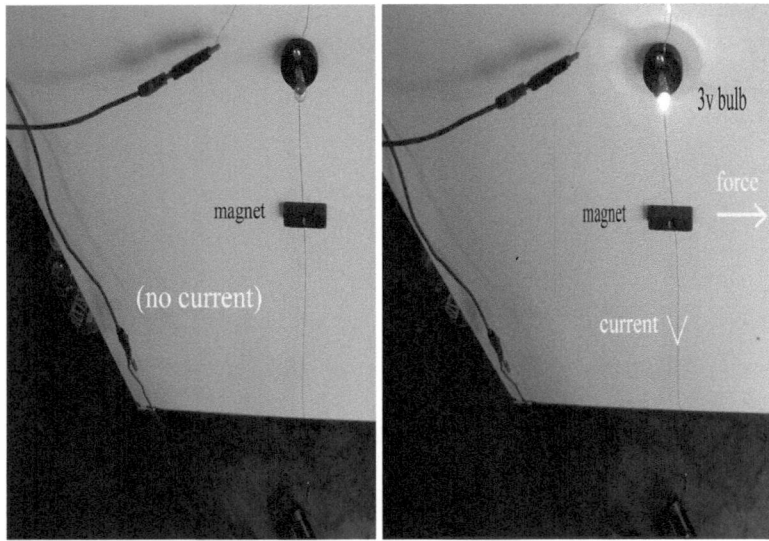

Fig. 3.20 View from higher up. In the *left photo* there is no current flowing (the two 1.5 V batteries, in series, may just be seen on the floor, beside a switch). When the current is switched on, the hanging wire is immediately forced to the *right*, because we have set the magnet with its field emerging directly out of the refrigerator door. (See also Fig. 3.21)

Fig. 3.21 The central finger of the right hand gives the direction the wire will move in. This is also referred to as the "corkscrew rule," because if you rotate from thumb to forefinger, the result is you go <u>in</u>

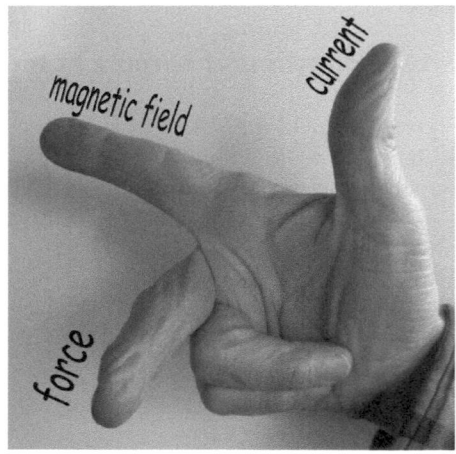

He was made Professor of Theoretical Physics at Leyden at 24 and, with Peter Zeeman, shared the 1902 Nobel Prize.

Lorentz is also known for the famous "Lorentz transformation" used in Einstein's relativity theory.

Nearly a quarter of a century after Lorentz' death Einstein wrote: *"For me personally, Lorentz meant more than all the others I have met on my life's journey."*

It's also interesting to note that the earliest electric motors, of the 1820s, -30s, and -40s, were actually developed before Lorentz was born.

3.17   A Kitchen Motor

First, if you prefer not to make our kitchen motor shown below, a clever, elegant, and absolutely minimal device is shown on the Web site *www.howtolou.com*—just click on "how to build a simple motor." That device is, however, quite remote in design from what you get in a cordless drill, for example, and the motor we describe here is much closer to the industrial motor—of which billions of examples have been made in the last 120 years (Fig. 3.22).

The motor to be constructed here is shown in the schematic of Fig. 3.23.

1. THE BASE, FRAME, MAGNETS, AND SPINDLE

   You will need the casing of a cheap ballpoint pen, internals removed, and a spindle, such as a nail, on which the casing will turn freely. They must be chosen to fit, and they define the dimensions of the motor. For the spindle we used a common 5″ nail, but a meat skewer, a knitting needle, or any other long straight object will serve. The pen casing needs to be cut to about 1″ shorter than the nail.

   You then need to make a rectangular steel frame to carry these items, as in Fig. 3.24. We found a product called "galvanized pipe strapping" which is suitable. (It must be steel to have magnetic properties.) If you are a user of an electric drill, and a hacksaw, and have any steel strip, then you can arrange something without advice from us!

   Our strip was bent into a rectangle approximately $4'' \times 3''$ with the cut ends meeting at the base, and it was fixed to a small plywood board by means of thumb tacks. Four tacks were hammered into the board, and these will locate the motor's brushes. The magnetic field is provided by small flat magnets (see p. 157). Four magnets are spaced along the base of the frame, with (say) their S poles upwards, and another four magnets spaced along the top of the frame, with their N poles downwards. With the steel frame this gives a powerful and fairly even magnetic field.

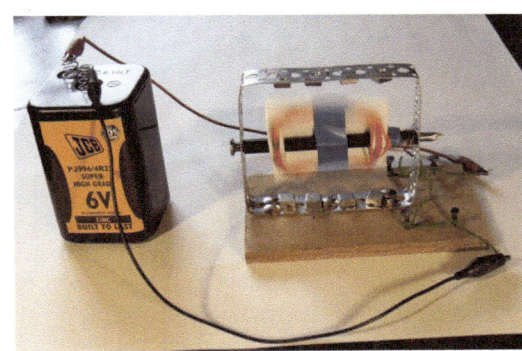

Fig. 3.22 Kitchen motor. (A smaller battery is also acceptable)

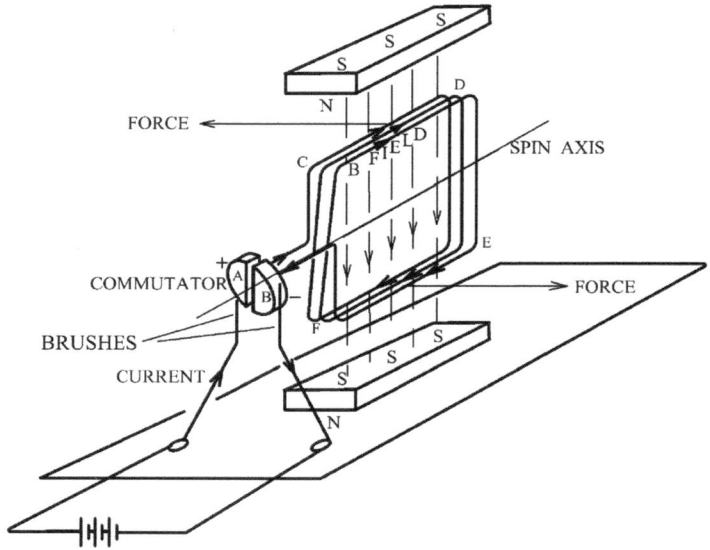

Fig. 3.23 Schematic for our home motor

Fig. 3.24 Frame, with
magnets *top* and *bottom*

2. THE ROTATING COIL

The coil is mounted on a light cardboard rectangle (cornflakes packet), which will
fit in the steel frame between the two rows of magnets with adequate clearance.
The cardboard has slits cut near each vertical side so that it can be threaded onto
the pen casing and held in place with sticky tape (blue in the picture).

Fig. 3.25 Coil on cardboard

Fig. 3.26 Making the coil

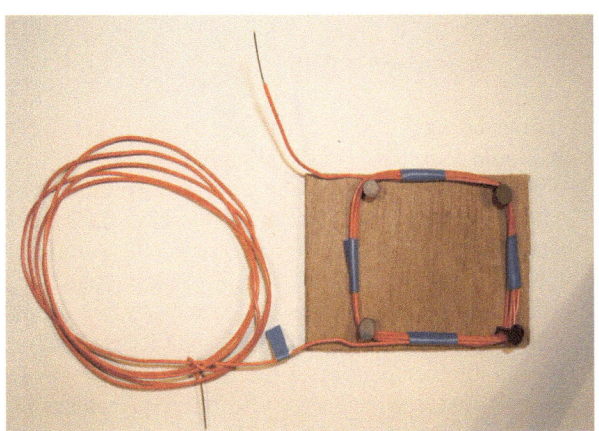

The coil itself will be wound from "hookup" wire, with a single solid conductor so that it is quite stiff. Our wire is orange in color, and the thickness of the core is 22 AWG or thereabouts.

The coil needs to be in two parts to fit either side of the cardboard, and this will be achieved using a separate plywood rectangle having four tacks spaced to form the corners of the coil, as in Fig. 3.26.

You will need about two yards of wire, stripping off ~1″ of insulation at each end. Find the mid point of the wire, and mark it with a little tag of sticky tape. Leave about 2″ at one end of the wire free, and then wind the wire around the tacks until you get to the tag, probably about six turns; stop, and bind each side of the rectangular coil with sticky tape to hold it all together (Fig. 3.26).

Now carry on winding in the same direction until you have wound the second half of the coil, leaving about 2″ free at the end. Bind three sides of the second half of the coil, leaving the side with the tag on it unbound. This is the section of the coil which is going to take the wire from one side of the cardboard to the other.

Fig. 3.27 The two parts of the coil

Extract one tack, and you will then be able to gently remove the coil from the board. It should look like this (see Fig. 3.27):

Now slip the coil over the cardboard former, and secure it with sticky tape. The free ends, about 2″ long, need to have their bare ends bent round in a hairpin shape. This will, surprisingly, serve as one sector of the commutator[2] which feeds the current into and out of the coil. Secure them in place with a thin strip of tape. Make sure that the hairpins are located at 90° to the plane of the coil as in Fig. 3.25.

3. ASSEMBLY, ADJUSTMENT, AND RUNNING

Here's another picture of the frame, but we have added two short lengths of green hookup wire. Each end of these pieces has its ends stripped for about ½″ and is anchored to a pair of tacks; they are going to act as the battery connections and commutator brushes. Notice how the upper bare ends of these wires cross close to where the spindle will pass through the steel frame (the hole is outlined in green) (Fig. 3.28).

Now maneuver the coil into position so that the commutator hairpins are pressing the "brushes" apart, and insert the spindle so that the coil is located correctly. It should look like this (see Fig. 3.29):

When you're happy that it looks like the above, set the coil to be vertical, check that both brushes appear to be making contact, and connect the battery to the green brush wires with your crocodile/alligator clips. At this point, if you're lucky, the coil will start to spin; but owing to the crudity of the commutator, it may merely twitch, in which case give it a flick in the direction of twitching; or, if this doesn't work, try a flick in the opposite direction.

---

[2] A device for changing the direction of the current through the coil every half turn. Without it, the coil would simply move half a turn and then stop!

Fig. 3.28 Showing the commutator brushes

Fig. 3.29 Complete assembly

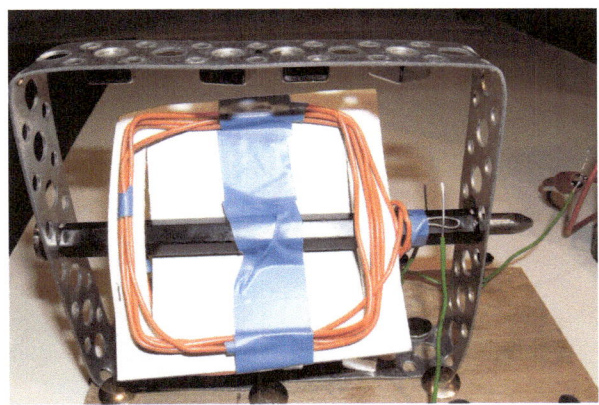

As you can see from the action photo in Fig. 3.22, it will run in a most entertaining way. (OK, we used a long exposure!) Also, in the true spirit of kitchen science, a smear of cooking oil on the spindle will allow it to spin faster still!

Exercise: Think through how the lowest section of the coil shown in Fig. 3.23 will begin to move, using the RH rule. Note: By convention the magnetic field is always from N to S, and so it is here DOWN. The (+) side of the battery (that's the longer line of the battery symbol) is shown to be going to the top side of the coil, which is labeled CD, and the current will continue along the bottom side of the coil from E towards F.

Further discussion: The force on the bottom side of a single coil is opposite to the force on the top side, because their currents are opposite.

By winding many turns of wire on a coil the forces are multiplied. Also, when a coil is horizontal the forces are cancelling each other, and if the coil were to run on a little bit, the forces would reverse, and the motor wouldn't go!

This was why we needed what we referred to as the *commutator*; it became necessary to reverse the connections to the coil at the crucial moment! The commutator comprises two semicircular segments A and B with an insulated gap between them, which are fixed to the coil and rotate with it. (In Fig. 3.23 A is connected to the C end and B to the F end.) Current is fed into A and out of B by *brushes* which are held against the segments by a certain amount of springiness. (They were originally bristles made of brass wire, but are now usually carbon blocks with a spring behind them.) As the coil reaches the horizontal position, A loses contact with the (+) brush, as does B with the (−) one. Momentum carries the rotation forward, so that almost immediately B makes contact with the (+) brush.

An obvious improvement is to have several coils arranged around the spin axis of the motor, each with its pair of segments—and the commutator is divided into numerous sections correspondingly, as can be seen if a professionally made motor is taken apart. Our motor (Fig. 3.22) may also be seen running at http://www.springerimages.com/videos/978-3-319-05304-2. (Search for the video "Making a motor").

Finally, the reader may be wondering "haven't we heard of *brushless* DC motors"? There are indeed such things, where the magnets are on the outside and rotate about stationary coils, but this is achieved by converting the DC, electronically, into AC pulses.

3.18   Adjacent Currents

Instead of bringing a magnet up to a current-carrying wire as in Fig. 3.19 and/or Fig. 3.20, we could simply bring up another current-carrying wire, because the current in the second wire is of course producing its own magnetic field.

For parallel wires the **B** field from one wire is perpendicular to the current in the other wire, as shown in Fig. 3.30.

Although this force is rather too weak to demonstrate at home, it is certainly true that the adjacent wires of a solenoid (for example) are slightly attracted to each other, such that solenoids tend to "shrink." Conversely, opposite currents will repel, easily verified again from the RIGHT-HAND rule of Fig. 3.21.

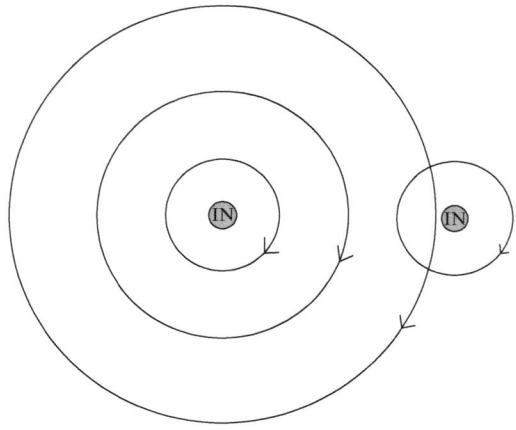

Fig. 3.30 Two adjacent wires. Here, both the currents are INTO the page. The force on the *right-hand* wire must therefore be to the LEFT. (*Thumb* IN; *forefinger* DOWN.) Thus, attraction

Fig. 3.31 When there's a
strong current, the coil will
contract slightly, thus
breaking the circuit at point A.
(Note that this is initially a
short circuit—things will
rapidly get hot if, by mistake,
contact perseveres!)

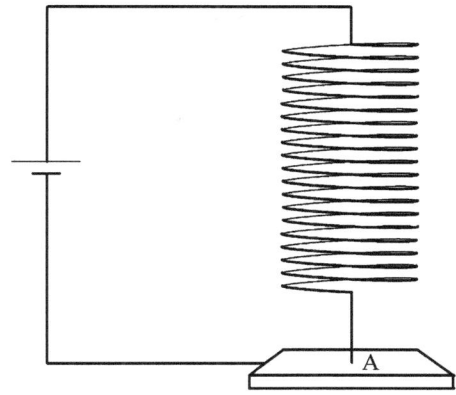

3.18.1 A Gedanken Experiment

In Fig. 3.31 a helix such as a "slinky" is just touching a metal plate at A, and so the short-circuit current flowing will cause the slinky to contract. (We have just seen that parallel currents going in the same direction attract.)

This contraction will break the circuit of course, and the slinky will then recover itself and retouch at A, and so on. In practice it is very difficult to achieve this, for the experiment is delicate and finicky!

For the next experiment, we will notice that it isn't just currents that are affected by magnetic fields, but *moving charges in general*, because currents are (as we emphasized in Chap. 2) nothing other than moving charges.

3.19   Lorentz Force with Old TVs (Not New Ones!)

In the twentieth century essentially all household TVs had CRTs or cathode ray tubes, but there are few left nowadays. In these old CRTs electrons came from a hot filament in the back of the tube and hit the front of the phosphorescent screen to give the picture. (The glass at the front was painted on the inside with phosphors, and when the electrons hit the phosphors light was emitted.)

A very easy experiment is to bring up a strong kitchen magnet to the side of (or above) an old TV and see the picture move respectively up or down (or sideways) (see Fig. 3.32).

CAUTION: Do NOT attempt this on modern TVs or computer monitors; they no longer use CRTs, but LCDs and/or LEDs, and the TV will be damaged.

Fig. 3.32 Lorentz force again: bending a beam of electrons in an old cathode ray tube TV. In this diagram, the first finger (field) is *"brown,"* the thumb (motion of (+) charges) is opposite to *"blue"* because electrons are negative, and so the beam is forced DOWN into the page

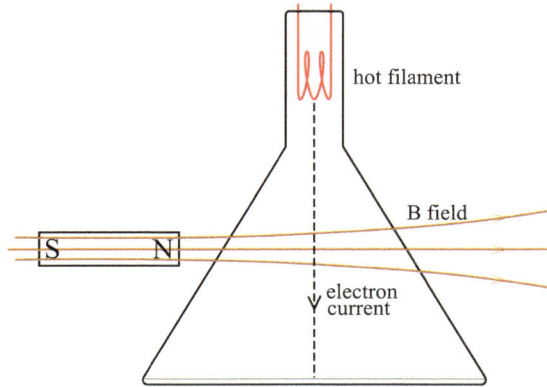

hot filament

B field

S   N

electron
current

3.20   Hall Effect

This same idea, that charges experience a force, is seen in the "Hall effect," which was discovered by the American, **Edwin Hall** (1855–1938).

Hall was born in Maine and discovered the effect in 1879—the year Einstein was born—while working as a graduate student at Johns Hopkins in Baltimore. He was later a professor at Harvard and the author of many physics texts and lab books.

He found that a magnetic field on a conductor would cause a "sideways" voltage, as shown in Fig. 3.33.

In the figure, the grey dots represent (+) charges, which would[3] initially travel (IF they were free to move) towards the yellow side because of the battery. But when an EXTERNAL magnetic field is brought near, those (+)s (again, if they were free to move) would travel to the side opposite the blue. However, the actually free charge carriers, the electrons, are the ones that move, and they go to the blue side. A potential difference is thus set up across the sides of the block, because electric charges are accumulating. This voltage is measurable.

Such a migration of charges is not surprising, of course, because of the Lorentz force.

The Hall phenomenon is widely used, more than a century after its discovery, for detecting magnetic fields. For example, if a change in magnetism occurs near to a Hall sensor, a buzzer could go off, and this could be used in metal detectors on the beach, etc.

Note that the force on the charges, or current, was given by the RIGHT-HAND RULE of Fig. 3.21.

---

[3] As we have repeatedly said, the (+)s are generally not free to move, so it's the electrons that do the actual moving. We have only drawn it this way for ease of application of the Lorentz force.

Fig. 3.33 A block of metal (Hall used gold), and a battery connected across it (with a resistor to limit the current)

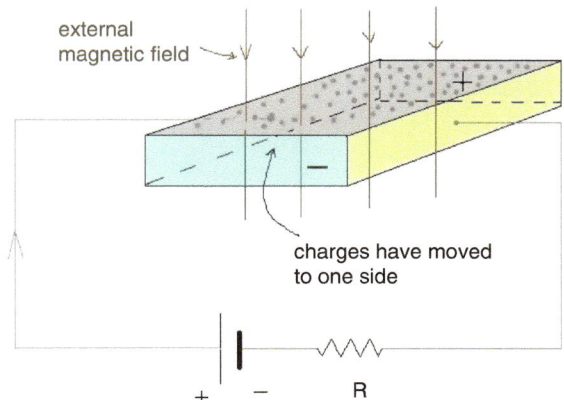

Incidentally, this "Hall voltage" could be calibrated and thus give a measure of the strength of the external magnetic field. We might then call the thing a "gauss meter." ("Gauss" is an older unit for magnetic field **B**.)

Such a voltage is too small to measure in the kitchen. However, one can buy an elementary Hall sensor (with a built-in amplifier) from the Internet for a few dollars. If this is done remember that, when making the connections, the Hall sensor is merely a slab of metal as in Fig. 3.33.

3.21   Magnetohydrodynamics (or MHD)

Magnetohydrodynamics is just a fancy word for another utilization of the Lorentz force.

The most important name in MHD studies is the Swedish scientist **Hannes Alfven** (**1908–1995**) who won the 1970 Nobel Prize for his lifetime of research into plasma physics and astrophysics, and most of his work was done alternating between Uppsala (Sweden) and California.

In Fig. 3.34, a glass of water is balanced on a neodymium magnet. There are two copper wires or electrodes: one at the edge and the other at the center. The water will become conducting if we pour some salt in it. We can then connect a battery (or even two batteries in series) to the electrodes, and the direction of the current doesn't matter.

Visualize this current along a radius in the water; it will experience the sideways Lorentz force of Fig. 3.21.

If we pour in some bread (or cookie) crumbs, or something like *turmeric* from the kitchen shelves, we'll actually see the fluid motion—clockwise or anticlockwise, depending on which way round the battery was connected. A short video of this home demonstration may be seen on http://www.springerimages.com/videos/978-3-319-05304-2. (Search for the video "MHD demonstration").

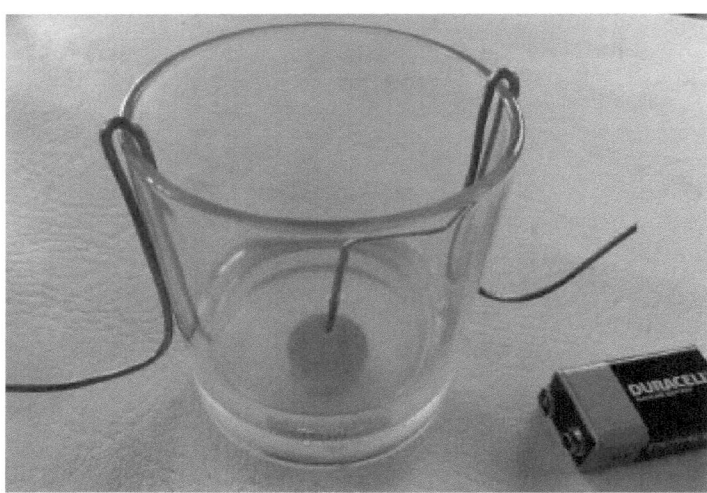

Fig. 3.34 A glass with an inch or so of salty water sits on a magnet. One copper wire is at the *center* and the other at the *edge*. When we connect a battery, either polarity, current will flow through the water

In sum, we have a B field (*magneto*), a fluid (*hydro*), and motion (*dynamics*)— hence the fancy title.

Note: Magnetohydrodynamics has been suggested as a means of propelling a seagoing vessel. If the water moves, then of course there must be an equal and opposite reaction. In principle, therefore, our glass-and-affixed-copper electrodes should recoil. For a boat the idea is that the salt water will move backwards and the boat forwards. Of course there wouldn't be much of a force at all in the relatively pure water of a lake!

In our kitchen experiment bubbles will be seen coming off each copper wire, especially if we used two 9 V batteries in series, and we alluded to this in our discussion of the composition of water in Chap. 1. Here the water has decomposed into ions, so hydrogen can be seen bubbling off the (+) electrode and oxygen off the (−) electrode.

The term used to describe this general decomposition into hydrogen and oxygen is, as we saw before, *electrolysis*.

3.22   Note on Microwave Ovens: A Subtle Example of the Lorentz Force

During WW2 (1939–1945) radar had been developed, and radar is nothing more than the transmitting and reflecting of radio waves. Radio waves are just a part of nature's *electromagnetic spectrum*, to be discussed further on pp. 114–115.

Our story is as follows. In 1946 a Maine-born orphan who had never had the opportunity to go to college, **Percy Spencer** (1894–1970) was an employee of the Raytheon Company. While working with radar waves he noticed that a chocolate bar in his pocket had partially melted. It was soon realized that instant cooking could take place at roughly the wavelengths being used, and the upshot was the AMANA "Radarange"—a massive 1950s microwave oven that cost about $5000!

Now there are many everyday examples of radio waves in this approximate region of the electromagnetic spectrum, given on p. 115, from the radio waves of cell phones to a policeman's radar "gun." The latter, for example, produces a beam of radio waves that can be aimed at a speeding car, and the waves are partially reflected back towards the "gun" where there is also a receiver. If the reflected waves have a slight shift in frequency (called a Doppler shift) then the speed of the car can be obtained. Also, in principle, the waves from a cell phone being used inside a speeding car must also be slightly shifted, but because the shift is so small this has not been known to cause problems.

So far, our discussion seems to have nothing to do with the Lorentz force. However, where do microwaves come from? In general, radio waves come from accelerated charges. It is an interesting and important result in physics that <u>any time charges are accelerated then electromagnetic radiation is produced.</u>

In the microwave cooker the waves are produced by something called a *magnetron*—a fairly easily replaceable part—sketched in Fig. 3.35. Here the electrons are again forced sideways, just as in the previous sections, thanks to the Lorentz force, and we have shown their approximate paths at the center of the diagram.

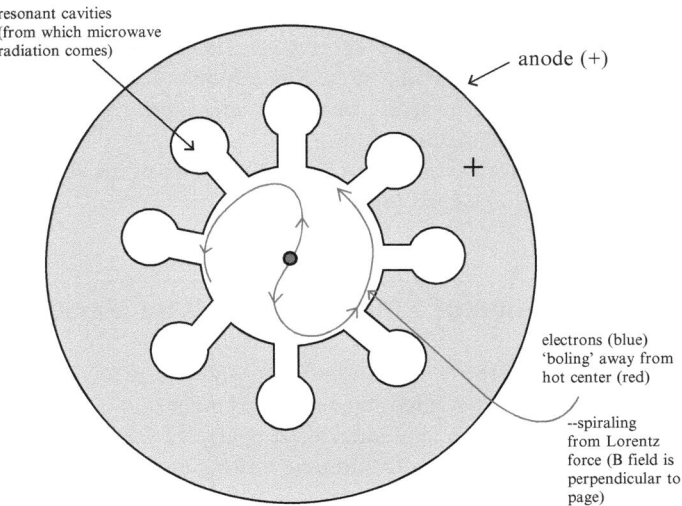

Fig. 3.35   The heart of the microwave oven is the magnetron. Just as there was a Lorentz force on a current, so now there is a Lorentz force on the electrons, which "boil off" from the central hot filament. Instead of going in *straight lines* to the anode they are deflected by a magnetic field (not shown here, but coming up from the page, parallel to the central filament)

What happens is that the electrons coming off a hot filament are bent into tight circles[4] and are thus accelerating.

It would involve too much of an advanced discussion to pursue further how radio waves come from the circular cavities of a magnetron, and the reader wishing to know more should research "cavity radiation" in general.

3.23 A Kitchen Experiment with Microwaves

Place a small glass of water and a small glass of oil (we used vegetable oil) side by side in a microwave oven, and switch on for about a minute. The water gets hot, and the oil much less so.

This is because, as we saw in Chap. 1, water molecules are *dipolar*, but oil molecules are not. As the radar waves pass a water molecule the two ends (or poles, + and −) of the molecule—as illustrated in Fig. 1.10 [water dipole]—vibrate back and forth, *resonating* in sympathy with the waves, and this can only happen at exactly that correct wavelength, **12.2** cm (*i.e.*, about **5** inches).

Such a rapid vibration of the water molecules is nothing other than heat.

Other foods experience some heating: fats and sugars absorb the radiation a certain amount, but the plastic and ceramic containers do not. The latter allow the electromagnetic microwaves to pass through unhindered.

Note that microwaves have nothing to do with radioactivity! Radar waves are adjacent in the electromagnetic spectrum to infrared, which are in turn (as their name implies) adjacent to the red colors. These wavelengths (*i.e.*, at the top end of the spectrum table of Fig. 3.46) are all relatively long wavelengths and of much lower energy than the waves at the shortwave end—which are the dangerous ultraviolet rays, X-rays, and gamma rays. Since humans are also mostly water, microwave ovens must have no leaks, of course, and there are strict government regulations on this for manufacturers.

We will return to radio waves after a brief discussion of an absolutely fundamental law in physics—*Faraday's law*.

3.24 Relative Motion of a Magnet and a Wire (Faraday's Law)

Michael Faraday (1791–1867) was a Londoner who had received only minimal schooling before a 7-year apprenticeship as a bookbinder.

After his apprenticeship Faraday attended a course of lectures by the chemist Humphrey Davy at London's Royal Institution, writing up very clear notes. He was

---

[4] Note that anything changing its speed and/or direction is indeed accelerating. For motion in a circle, one feels such acceleration in any merry-go-round. Instead of flying off in a straight line we are continually being forced towards the center. Actually, if we were <u>electrically charged</u> we would emit electromagnetic radiation!

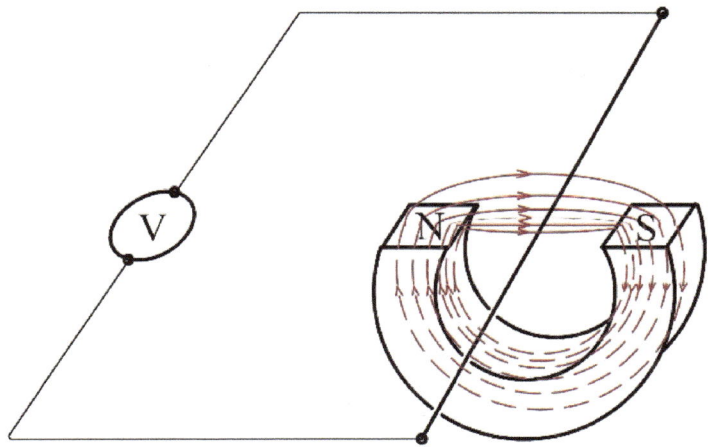

Fig. 3.36 If the wire (*black*) is moved up and down in the magnetic field (*brown*), a tiny alternating voltage will appear across the ends of the wire

consequently employed as an assistant, working in chemistry, electricity, and magnetism, and one cannot here begin to describe all his discoveries and his ultimate worldwide fame.

One of his discoveries was that if a magnet is moved near a wire or a wire near a magnet (it doesn't matter which) there will occur a voltage across the ends of the wire! This discovery is referred to as **Faraday's law** of electromagnetic induction.

Faraday's basic demonstration would be as in Fig. 3.36, if we had a strong horseshoe magnet such as often found in physics labs.

However, we <u>can</u> easily do the experiment in the kitchen, as in Fig. 3.37, where we have resorted to an inexpensive general household meter (see p. 159). Such common and useful meters can measure voltages, currents, and resistance.

The most sensitive current scale for this kind of meter is <u>250 *micro* Amps</u> or 0.25 mA—very small indeed, and fortunately many equivalent meters on the market today have such sensitive scales.

<u>Note</u>: **<u>Be careful to keep strong kitchen magnets and the current meter well apart—a good three feet. It could wreck the meter if any strong magnet is brought too close</u>**. Luckily our meter's red and black leads here are about 4 feet long.

All we do is move the magnet (left photo) over the coil—the more turns in the coil the merrier—back and forth, rapidly. Small readings of a few microamps will be observed on the meter, and as long as the back-and-forth readings remain small, it doesn't matter that the needle oscillates the "wrong" way, *i.e.*, below the zero mark.

What was important in the above experiment was the *relative* motion of wire and magnet. It can also be seen that the *faster* the motion the greater the effect.

These results are summed up in Faraday's Law: "A changing magnetic field will induce a voltage (called an e.m.f. or electro-motive force) in a conductor nearby" or, equivalently, "relative motion between a magnetic field and a conductor will induce a voltage in the conductor."

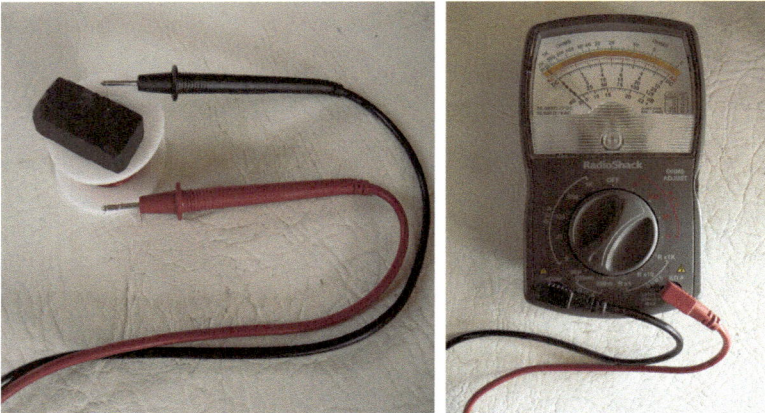

Fig. 3.37 *Left photo*: Magnet and coil. *Right photo*: All-purpose multimeter (set on 250 microamp scale)

It is important for us to point out that Lorentz's law and Faraday's law are experimental converses, in the sense that Lorentz's law gives us the electric motor and Faraday's law gives us the generator.

3.24.1   Note on "e.m.f.s"

One often comes across the abbreviation *"e.m.f."* standing for *"electromotive force"* (NOT electromagnetic field, although that confusion will too often be found in newspaper and magazine articles).

Electromotive force is simply a voltage—the voltage before current is drawn.

The distinction may be seen by considering a battery. The nominal voltage (e.m.f.) of a D or a C or AA or AAA cell may be 1.5 V, but as soon as it is connected it drops down a little, perhaps to 1.4 or 1.3 V, depending on how much current is flowing.

For this slight reason, "e.m.f." has hardly been used in place of voltage or potential difference in this book.

3.25   Transformers: An Example of Faraday's Law

Evidently, what we have just seen in Figs. 3.36 and 3.37 gave us a little voltage because of relative motion between wire and B field. Another way of producing this effect would be to cause the B field to fluctuate around the wire. If we set up two adjacent coils, as in the left side of Fig. 3.38, and a little pulse is given to the bottom coil (the primary) from a brief connection to a battery, then there will be a brief change of magnetic field, from zero to some finite value, in a very short time—and if the top coil (the secondary) is close enough it will feel this change. A tiny momentary voltage, or kick on a meter, will be detected on the upper coil. We call this setup a transformer, as we will see in the example on the next page.

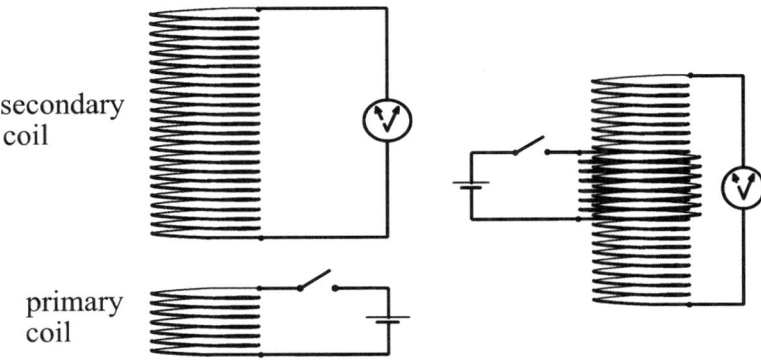

Fig. 3.38 The reader should visualize the sudden **B** field created on the *left* by the primary when the switch is thrown momentarily. If the secondary is close enough it will feel the sudden field and give a little kick on the meter. The setup on the *right* will be more efficient, especially if there is an iron core (see also Fig. 3.16)

We have intentionally omitted the B field lines in Fig. 3.38 so as to avoid clutter. The transformer will be better if the coils are as shown on the right side of Fig. 3.38, together with an iron core, because then the B field is strong, and common to each, with much less leakage, as in the earlier Fig. 3.16 on p. 90.

Faraday described the B field as a magnetic *flux* (Latin, *fluere*, to flow) because its shape resembles the flow patterns you get in a liquid. (The actual definition of flux is *field multiplied by area*, as defined in the Appendix on p. 169.) He realized the significance of both the <u>strength</u> of the B field and the <u>cross-sectional area</u> of the field which is interacting with the wire or the coil, as in the electric motor.

Inside the coils on the right of the above figure the field (and/or flux) is nicely concentrated in the interior—as we sketched earlier in Fig. 3.16.

Transformers only work with <u>changing</u> fields, as Faraday's law states, so with DC you will only get a fleeting initial "kick" when the switch on the primary is closed (or opened). If an ongoing AC voltage is applied to the primary, an ongoing AC voltage appears on the secondary.

The magnetic field produced by the primary coil depends on the number of turns, N_p, and to get a stronger field we could increase this. Likewise, the voltage induced in the secondary coil also depends on its number of turns, N_s. If the field is pretty much common to each, as in the right hand of Fig. 3.38, then the ratio of these two numbers of turns, *i.e.*, N_s/N_p, will be the ratio of the voltages, as in the following example.

Exercise: On the left side of Fig. 3.38 an AC voltage of 10 V is applied. What will be the secondary voltage?

Answer: Let us assume that there are 1200 turns on the secondary and 300 on the primary. The voltage ratio will be 12:3 or 4:1. This means that we'll get 40 Volts on the secondary.

Also, transformers may either step up or step down, so if we had put the 10 Volts on the secondary, we would only get 2.5 Volts on the primary.

3.26   Two Examples of Transformers

1. <u>Large:</u> In the USA, the high voltage coming down the local road is typically 7620 V, so this must be transformed down to something more friendly.

   Sometimes the arrangement in a transformer is as in Fig. 3.39, where the secondary winding has a center tap, thus yielding <u>two</u> voltage sources. In a typical US house for example, most items like lamps, TVs, radios, and clocks run on 120 V, but a supply of 240 V is used for larger items, like the water heater or the clothes dryer, by connecting directly across the two 120s. Such a transformer is shown in Fig. 3.40 and is most likely to be seen on a pole on the road not far from a house.

   This grey cylindrical looking object changes 7620 V AC[5] down to a safer 120-0-120 V AC. (In other countries these values may be different.)

2. <u>Small:</u> Little transformers for telephones, chargers, and so on, as in Fig. 3.41, typically supply about 5 or 6 V AC, and this may be tested directly on the appropriate AC voltage setting of a multimeter. (It is suggested the 20 V AC setting be tried first.)

   Astonishingly, our multimeter may give BOTH AC and DC readings for the little transformer shown, as illustrated in the readings of Figs. 3.42 and 3.43. Look at the white dot in each of Figs. 3.42 and 3.43.

   On the AC photo the dot is on the AC setting, and on the other photo it is on the DC setting. This can only mean that we do indeed have both.

Exercise: How can this be?

Answer: We know that a transformer can only change AC to AC. The waveform we are getting is as illustrated earlier, on the right-hand side of Fig. 2.42 on p. 73.

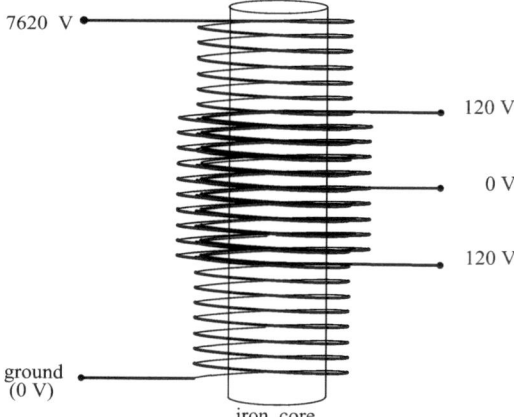

Fig. 3.39 The secondary, on the *right*, gives two sources of 120 V

---

[5] Information from Central Hudson Gas & Electricity, Poughkeepsie, NY.

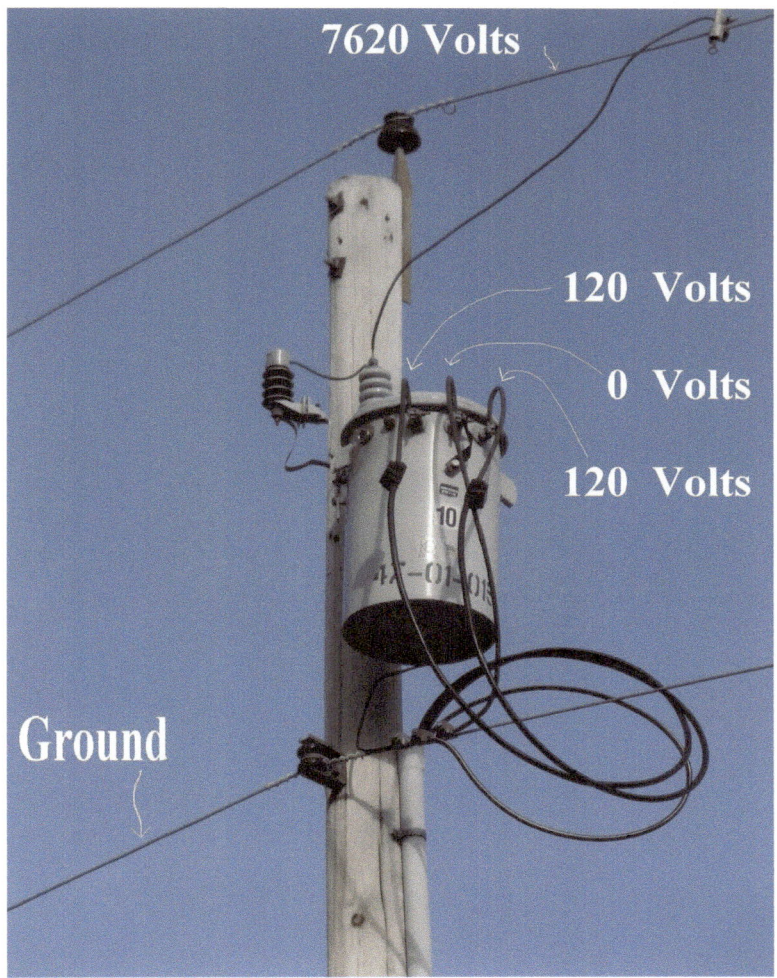

7620 Volts

120 Volts

0 Volts

120 Volts

Ground

Fig. 3.40 A typical step-down transformer in the USA, showing what the wires are. The house supply of 120-0-120 goes down a conduit at the *bottom* of the photograph. The center tap is connected to the ground wire

However, if we were to cut off the bottom half of an AC waveform, by means of a diode, then we would have DC also—albeit a rather jagged DC. Our meter can thus read <u>both,</u> and you can also see that the AC and DC voltages are slightly different.

Fig. 3.41   Just above "NOM" it says that the output is 5.9 V <u>DC</u>!!! This means that there must be a diode somewhere after the secondary winding of the transformer

Fig. 3.42   Showing AC Volts

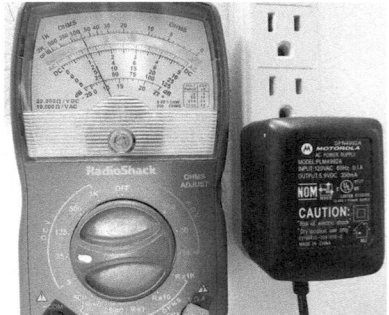

Fig. 3.43   Showing DC Volts

3.27   Electromagnetic Waves

We now know enough kitchen electricity and magnetism to describe a radio wave, and we illustrate it in Fig. 3.44, which depicts a wave traveling towards a straight wire antenna.

We saw in an exercise in Chap. 1 (p. 18) that an electric charge has a "field of influence" around it, emanating from it; physicists call it the **E** field. This field depicts the path that a tiny free charged particle, or test particle, will be influenced to travel along.

Although we didn't mention this, that **E** field actually emanates at the speed of light. We can't see or touch this field, but it is nevertheless a useful idea.

All electric charges come with this associated **E** field. If we were to take a charged object, such as a comb, and shake it up and down, the emanating electric field would go up and down, while propagating outwards, rather like the wave on a rope when we shake one end up and down. This is the blue line in Fig. 3.44.

We also know that the comb being shaken up and down constitutes a current, and from Oersted's experiment we must get a magnetic field, with direction here shown in brown. The blue and the brown are inseparable—they must both exist. This is an electromagnetic wave, propagating out at the speed of light.

Radio, TV, microwaves, visible light, ultraviolet, X-rays, and so on (see p. 115) all consist of these rapidly varying fields, as represented pictorially in Fig. 3.44. However, we only showed what is called a *polarized* wave—in this case polarized vertically. Also, it is conventional to describe only the **E** (blue) waves—where it is understood that the magnetic wave must always come along with it.

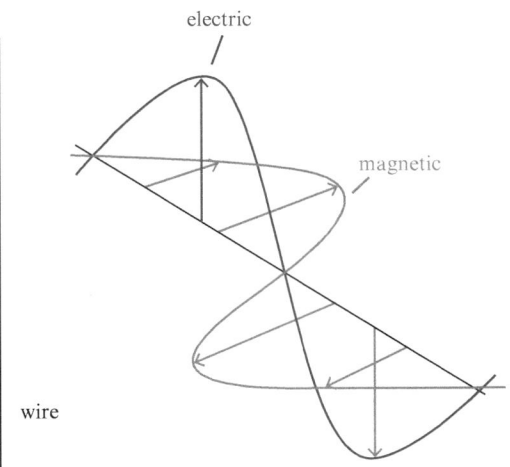

Fig. 3.44 A radio wave passing a wire. The plane of the magnetic field (*brown*) is always perpendicular to the plane of the electric field (*blue*) in any electromagnetic wave, and the fields are inseparable. We have shown only one pair, but for unpolarized waves they are in all planes

3.28   The Electromagnetic Spectrum

Sir Isaac Newton (1642–1727) noticed that the sunlight in his room could be split into all the colors of the rainbow by means of a prism, as in Fig. 3.45. What we now know, and he certainly didn't, is that beyond the red lie radio waves, and beyond the violet are X-rays and gamma rays.

All these radiations are electromagnetic. We give below some of their approximate wavelengths and frequencies, with an emphasis on the major radio parts.

Figure 3.45 includes something called "**RFID**" which stands for **R**adio **F**requency **ID**entification. These radio waves are commonly used when "tagging," whether it's store items or tolls for cars, etc. What happens is that radio waves are sent out from a transmitting base, and a nearby card ("tag") perhaps from a purchased item, or a car tag, sends back a similar but weak radio signal. Information (price, quantity, etc.) concerning the item can be recorded, without any need for touching or "swiping." These systems and methods are described further in the Appendix on p. 161.

The reader may notice that if the frequency and the wavelength, for any of these radiations, are multiplied together, one always gets the same number. This is because there is a general relationship for any kind of wave—ocean waves, sound waves, or others—which says that *speed is frequency multiplied by wavelength.*

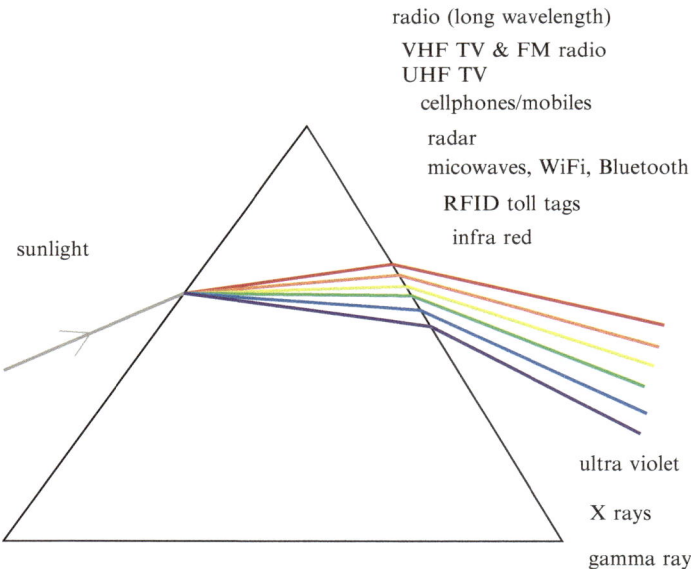

Fig. 3.45 White light can be split into all the colors of the rainbow. *Red* has a longer wavelength than *blue*. Beyond the *red* would lie the whole radio spectrum, and beyond the *blue* would be the very short wavelengths of UV rays, X-rays, and gamma rays (although not actually coming through the glass prism)

[Type of radiation]	*[wavelength]*	*[frequency]*
Radio, long wavelength	200 meters	1.5 MHz
some RF ID	22 meters	13.56 MHz
Radio, short wave	15 meters	20 MHz
TV VHF	3 meters	100 MHz
TV UHF	0.5 meters	600 MHz
Cell phones	0.4 meters	800 MHz
some RFID	0.3 meters	900 MHz
some toll tags	0.3 meters	915 MHz
GPS	0.25 meter	1 200 MHz
WiFi	0.12 meters	2 400 MHz
Microwave cookers	0.12 meters	2 500 MHz
Radar	0.07 meters	4 400 MHz
Infra-red (*i.e.* heat)	0.000 01 meters	30 000 GHz
Red light	0.000 000 7 meters	440 000 GHz
Blue light	0.000 000 4 meters	770 000 GHz
X rays	0.000 000 000 1 meters	3 000 000 000 GHz
Gamma rays	smaller than an atom!	30 000 000 000 GHz

Fig. 3.46 Some approximate wavelengths. In mankind's history, light was detected first, and radio waves were not discovered until the 1880s (see p. 73)

In Fig. 3.44 we showed exactly one wavelength. The number of times per second that a whole wavelength passes a wire (or an antenna) is called the *frequency*, and its definition has already been given on p. 73.

Note that

"MHz" means 1 000 000 Hz, and

"GHz" means 1 000 000 000 Hz.

3.29 Making a Kitchen Radio

We might wonder if <u>any</u> relative motions between fields and wires would cause voltages to be induced in <u>any</u> conductor. Might we not detect voltages from all around us—because even if we are standing still we are in an (ever-increasing) sea of passing radio waves. So, surely we will detect things? Cell phone users, of course, know the answer well!

What the previous sections imply is that if a length of wire lies in the path of any radio waves, as in Fig. 3.44, little voltages will indeed be induced in the wire. The free charges in the wire will go up and down, and this means that we have some alternating currents in the antenna.

If the wire in Fig. 3.44 is turned through 90 degrees so as to be parallel to the magnetic field (brown) rather than the electric field, then there'll be no induced voltage, because, as we saw in the previous sections, the magnetic field has to be at right angles to the conductor (see Fig. 3.36). However, we noted that the wave depicted in Fig. 3.44 is polarized in the plane of the paper. (In general, e-m waves are not polarized in any one plane—any and all planes are possible. For example, the light from an incandescent bulb is not polarized.)

So in principle, if we connect some headphones across the ends of an antenna, we should be able to detect voltages (actually modulated voltages) from radio stations. This energy, tiny but free, is in the radio waves themselves, paid for by the transmitting station owners. (One may imagine the owners rapidly vibrating charges or a magnet up and down—that's the original source of energy.) But that is only in principle—for we will actually need a diode to hear voices.

The reason is as follows: The diaphragm of a little earphone could in no way vibrate back and forth millions of times per second, and that is anyway well beyond our hearing range. However, if the radio wave has been modulated in some way by a regular sound wave impressed upon it then this slower variation (between ~20 and 20 000 Hz) should be detectable. The trouble is that the earphone diaphragm is pushed one way but at the same time the opposite way and so doesn't move.

However, if a diode is inserted then the earphone diaphragm would make an "audio excursion" in one direction only, albeit in varying amounts. We would thus have a "pulsating DC"; that is, we could hear the sound.

In the 1920s, "cat's whisker" (this was the diode) radios or crystal sets were common—and we can make such a classic radio now, as in Fig. 3.47. It is also worth looking up "crystal radios" on the Internet—there are literally hundreds of hobbyists who make these fun things and who give videos and photographs of their individual boxes and setups.

For the not faint of heart, what will be needed for a kitchen radio is ~50 foot long antenna outside—just plain wire; a homemade coil (about 120 turns of thin insulated wire on a cardboard former, such as from a plastic wrap roll); a variable capacitor, perhaps from an old radio; a laptop OR a ceramic earbud ($2.97); and a germanium diode (e.g., 1N60 or 1N34A, ~$1.50) instead of a "cat's whisker."

A germanium diode is better than a silicon one for such tiny voltages and certainly better than the *galena* used in the crystal sets of the 1920s.

This radio (Fig. 3.47) was able to receive shortwave AM stations at wavelengths of around 200 yards, which corresponds to a frequency of approximately 1200 kHz or 1.2 MHz. The actual signals received by our above radio may be heard by visiting the website extras.springer.com. On the website search box enter the book's ISBN number (978-3-319-05305-9) to find the file "Making_a_radio.mp3".

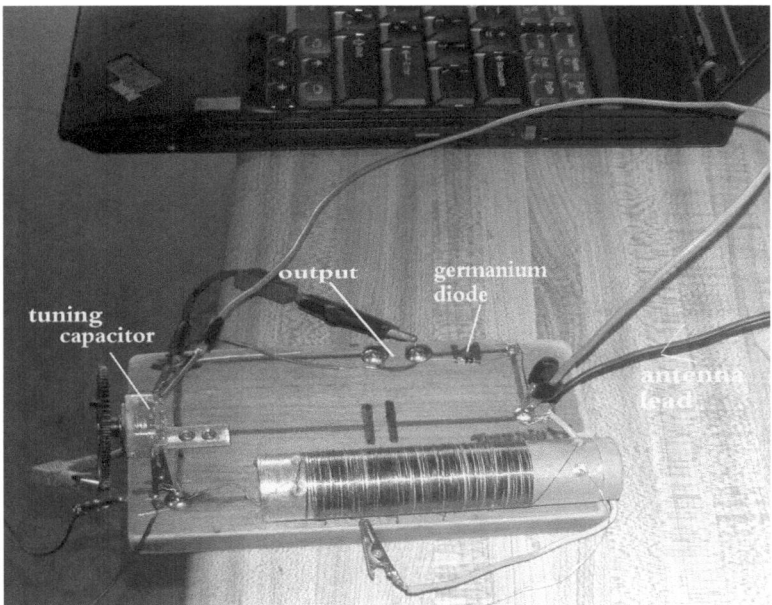

Fig. 3.47 The tiny voltages (AC) induced in the 50′ outside antenna are turned into DC by a diode, and signals (no batteries) can *just* be heard in the earbud. If they are too faint, the output may be fed straight into the microphone jack of a laptop and listened to over the laptop's speakers

3.30 Experiment: Falling Magnet

Another home experiment, much easier than the previous one, shows yet again how Faraday's *relative motion between a magnet and a conductor* can and will cause an induced current.

Take a nonmagnetic metal tube (*e.g.*, a section of vacuum cleaner pipe, not plastic, or a copper pipe), and drop a strong neodymium (see note on next page) magnet down inside it. Depending on the strength of the magnet it may take quite a while to fall! With the 1 foot section of copper pipe shown on the floor (Fig. 3.48) the time of descent for the little $8 cylindrical neodymium magnet was 3 s, and with the wider vacuum cleaner section it was about 1 s. This can be seen in http://www.springerimages.com/videos/978-3-319-05304-2. (Search for the video "Falling magnet").

3.30.1 What Was Happening

Faraday's law states that relative motion of a conductor and a magnetic field (as in Figs. 3.36 and 3.37) induces a voltage and a consequent current.

The falling magnet induces currents around the conducting tube, as labeled on the right-hand side of Fig. 3.48. This circulating current is essentially the same as the current circulating in a solenoid.

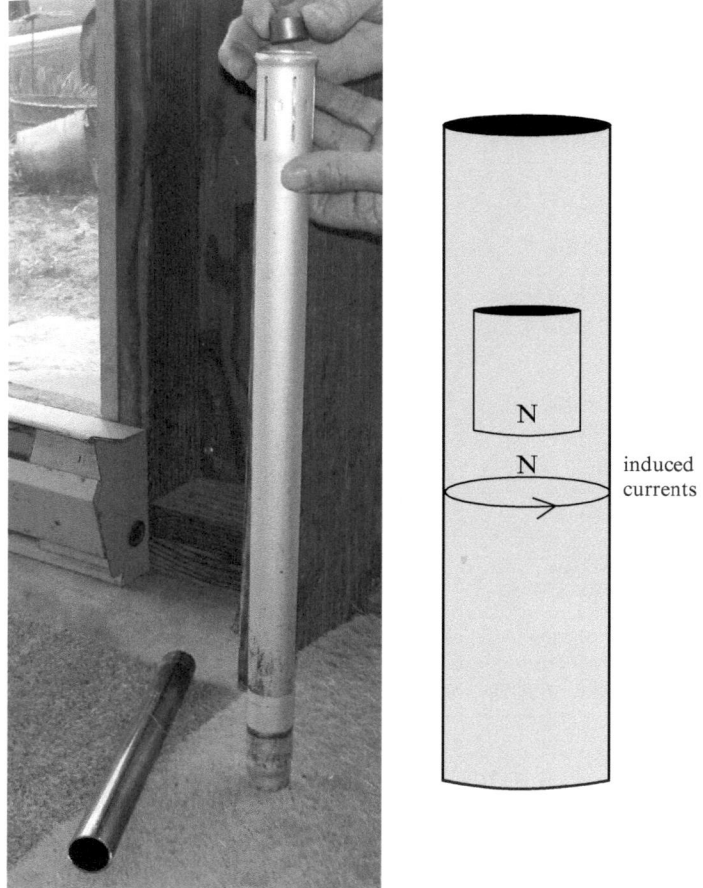

Fig. 3.48 A strong magnet (*e.g.*, neodymium) falling inside a conducting but nonmagnetic tube (copper or aluminum), taking a long time to reach the bottom. The induced currents cause a magnetic field which opposes the magnet's field

Let's say that the falling magnet has its "N" at the bottom end. Then the circulating current that is induced in the metal pipe gives rise to a field inside the tube (again, just like the solenoid field of Fig. 3.16) which has its "N" pointing *upward*, thus repelling the magnet and slowing the fall.

This is also an illustration of Lenz's law (**Heinrich Lenz**, 1804–1865—an Estonian-Russian physics professor at the University of St. Petersburg in Russia). Here his famous law says that "the induced current will always oppose the change that caused it"—but in fact it's a general law that occurs elsewhere in physics. Note that we have already met it when we described inductance (L); the induced voltage *opposed* the applied voltage.

3.30.2 Note on Neodymium Magnets

The strongest little permanent magnets used now in cordless tools, door fasteners, etc. are a compound of neodymium—iron—boron—and were developed in the 1980s by General Motors in collaboration with other companies. These magnets have now replaced Alnico and other older ferrite magnets.

Neodymium is a rare earth element (atomic number 60) and is mined in many countries, particularly China and the USA.

These neodymium-type magnets are not only extremely strong but also brittle. If they clash together too fast they can chip, or even shatter, and/or hurt your fingers. Also, because neodymium corrodes easily the magnets are usually coated when manufactured.

Google searches for these strong magnets change with the years. One of the authors obtained his astonishingly powerful little cylindrical magnet shown at the top of the left photo of Fig. 3.48 (diameter 7/8″ and depth 1/2″) for about \$16 a pair from Herbach & Rademan at www.herbach.com. To give an idea of their strength, the first mailing didn't arrive, so the two incredible magnets were probably clinging to something in the postal system! Fortunately, a substitute pair was sent.

3.31   Eddy Currents, ARAGO, and a Kitchen Cooker

In the early 1800s the scientist who would become France's 25th President (albeit only for a few months in 1848), **Dominique Francois Jean Arago (1786–1853)** noticed that the oscillations of a compass needle as it settled were slowed, or damped, if there was some copper underneath it.

In other words he noticed that there was some interaction between a magnet and a conductor IF there was relative motion, and, essentially, Arago's disk is going to turn out to be no different from what we've just seen.

One can see beautifully made "Arago disks" by looking up just that on the Internet. They are typically nicely made wooden boxes which contain a horizontal copper disk that rotates by means of a hand crank and with an ordinary compass just above (but not touching) the disk. If the compass needle is initially stationary, and the experimenter then cranks the handle, the compass needle will begin to rotate also. The disk and the compass are usually separated by a sheet of glass, so that there is no interference from moving air.

Arago couldn't explain the phenomenon, but a few years later Michael Faraday did.

Once again Faraday's law says that little currents must be induced in the copper or the aluminum disk from the relative motion between magnet and metal. These little circulating currents (he called them *eddy currents*) naturally produced their own little B fields—which then interacted with the permanent B of the compass needle.

As we emphasized, it doesn't matter which one moves—we only need *relative motion*. Thus, if we were to shake the compass needle back and forth, fast, this

would cause tiny eddy currents in the metal disk. The disk wouldn't have time to
rotate one way or the other, but the eddy currents would be there, and the metal
would in principle begin to get warm. Shake a neodymium magnet extremely fast
and one might be able to detect the warming in a close by sheet of metal, although
we have not done this. One would have to shake awfully fast, and if it did get warm
we would call this *induction heating*.

Today some kitchen ranges have a cool, shiny, ceramic surface, with high-
frequency oscillation coils underneath (instead of shaking magnets!). Manufacturers
use VERY-high-frequency oscillations in the coils of about 27 000 Hz (corresponding to a wavelength of about 6 miles) rather than the usual household current
of 60 Hz (or 50 Hz in Europe).

Induction cooking is very clean—there's no flame, and when you remove the
pan or the kettle the stove isn't dangerously hot—except for the fact that there was
direct contact with the now heated container. The eddy currents were in the kettle
itself—not in the stove top.

Induction cookers and ranges are sold in many appliance stores worldwide.
Fig. 3.49 shows a 1909 induction cooker (illustration taken from the public domain

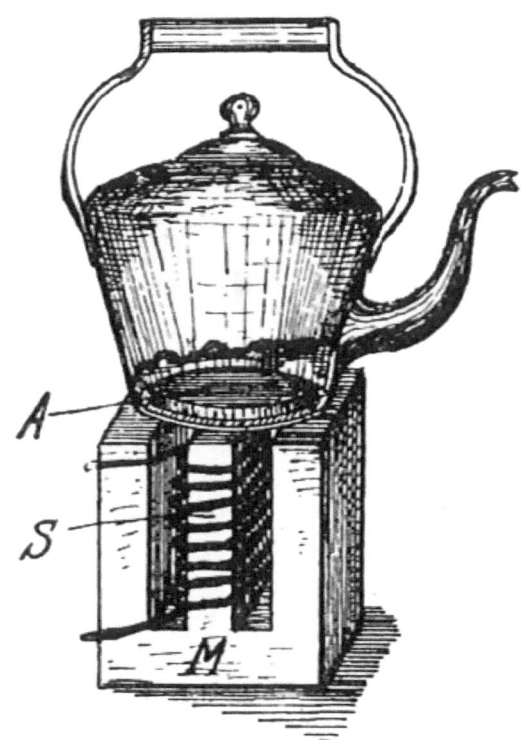

Fig. 3.49 Heating by eddy
currents (induction heating).
The eddy currents are induced
only in the base A of the kettle
or the pot

on the Internet). Here "**S**" is the coil that magnetizes and demagnetizes "**M**," and the rapidly changing magnetic field goes through the base "**A**" of the kettle, inducing eddy currents.

Note that the kettle material <u>must</u> have some resistivity in order to get the usual Joule heating that we introduced on pp. 16, 55. Also, although Arago didn't use magnetic materials such as cast iron, eddy currents will still form because iron is a conductor. Today, manufacturers of induction cookers sometimes provide their own pots and kettles, from their own alloys, having designed them for maximum heating.

3.32   A Household Generator

Finally, as we pointed out on p. 108, Faraday's law is the converse of Lorentz's law (which describes the principle underlying the electric motor). So if a motor is driven by a steam engine, say, then electric current will be generated. Another common example is the alternator driven by a car's engine.

In general, if we were to rotate a coil of wire continuously in one direction, in a magnetic field, we can generate useful electricity rather than random eddy currents and heat. This is what any power station does.

Figure 3.50 is a photo of a battery-less kitchen flashlight. The handle is squeezed back and forth, spinning a magnet in one direction, and this relative motion may well generate enough current to light a bulb.

The "gizzards" of this particular generator are shown in Fig. 3.51.

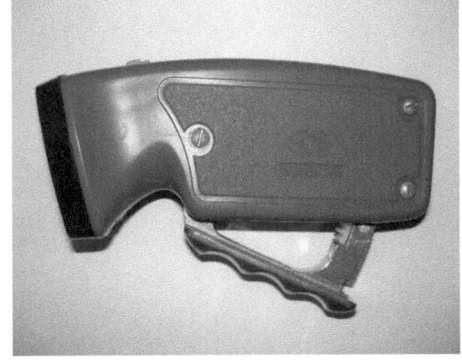

Fig. 3.50 By squeezing the handle vigorously, little magnets rotate, inducing sufficient current to light the bulb

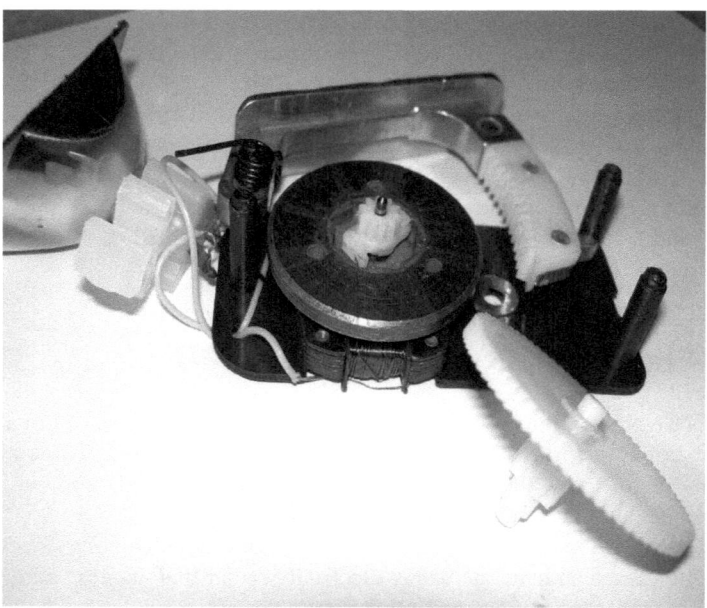

Fig. 3.51 The gizzards—showing the fine wire coils (there's a coil on the far side also) and the rotating magnet. (The latter are embedded in the under side of the metal disk and cannot be seen.) Again, all that is needed is relative motion between magnets and wire; and coils of wire are better than just one wire, of course

Elements of Transistors, and an Integrated Circuit

The *transistor* (the word comes from transfer and resistor and we shall soon see why) was invented and developed in the mid-twentieth century, and in 1956 the Nobel Prize for physics went to three men—Shockley, Bardeen, and Brattain—for their pioneering work on what became a common and cheap device.

John Bardeen (1908–1991) stands out, partly because he is a double Nobel Laureate in physics. His later Nobel Prize, in 1972, again shared with two others, was for work in superconductivity, which is that extraordinary effect where current can flow forever, without any resistance, at very low temperatures. However, superconductivity cannot be done in the kitchen!

So what is this little component that today runs everything from TVs and radios to automobile circuitry and cell phones? And which is so tiny that many thousands of them could fit onto something the size of a thumbnail?

Short answer: It is a switch, and/or it is an amplifier.

4.1 First We Must Revisit the Diode!

We have to look deeper into the diode of Chap. 2 because otherwise we will be rendered unable to understand the transistor.

We already know the difference between conductors and insulators: conductors such as copper and aluminum have low resistivity and insulators very high, as in the table on the next page. From that table we see that silicon seems to be an insulator, but we'll see that this depends strongly on whether impurities have got into it or not. Even if relatively pure it's not a strong insulator like glass, but neither is it a conductor, so it is called a *semiconductor*.

The online version of this article (doi:10.1007/978-3-319-05305-9_4) contains supplementary material. This video is also available to watch on http://www.springerimages.com/videos/978-3-319-05304-2. Please search for the video by the article title.

<u>Resistivity table</u> (approximate values, given in Ohms × meters × 10^{-8} at room temperature)

Silver	1.6
Copper	1.7
Gold	2.4
Aluminum	2.8
Mercury	10
Iron	11
Tin	11.1
Carbon	300
Seawater	~20 000 000
Germanium	~46 000 000
Silicon	~50 000 000 000
Polyethylene	10 000 000 000 000 000
Porcelain	15 000 000 000 000 000 000
Glass	1000 000 000 000 000 000 000

Silicon, and other elements such as germanium, can be made to become considerably more conducting <u>if</u> they are *doped* with (*i.e.*, mixed with) other atoms.

Because a picture is worth a thousand words, look at Fig. 4.1—where a red circle has been drawn around an ordinary silicon atom. For simplicity only, the diagram is shown in two dimensions rather than three.

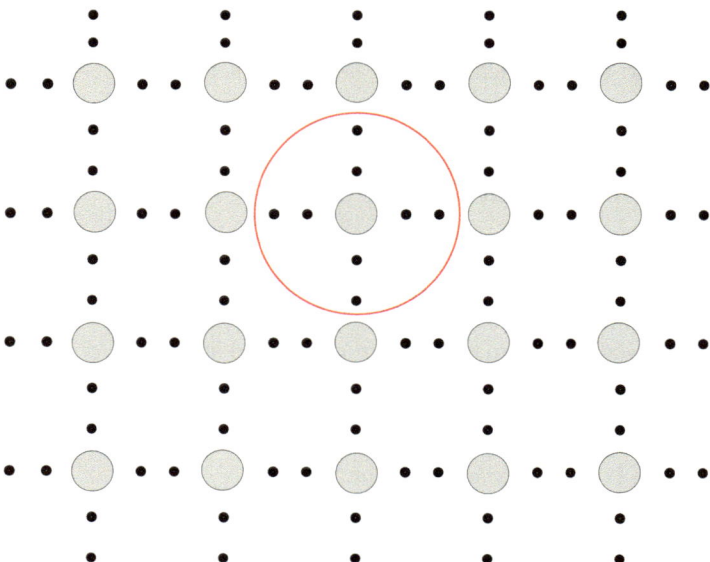

Fig. 4.1 Pure silicon atoms

From the periodic table (see any source containing the periodic table) an atom of silicon (atomic number 14) has a total of 14 electrons, and therefore 14 protons in its nucleus. As we saw in Chap. 1 the innermost shell has two electrons (not shown in our Fig. 4.1) and the next shell has eight (also not shown.) That leaves four electrons in the third and outermost shell (shown below).

As we said, pure silicon is nonconducting. However, if a "foreign" atom with (say) *five* electrons in its outer shell (an example would be arsenic) is introduced into an array of silicon atoms, four of those electrons will neatly "fill" the silicon's outer shell, leaving the fifth one as a *wandering electron*, like this green one shown in Fig. 4.2.

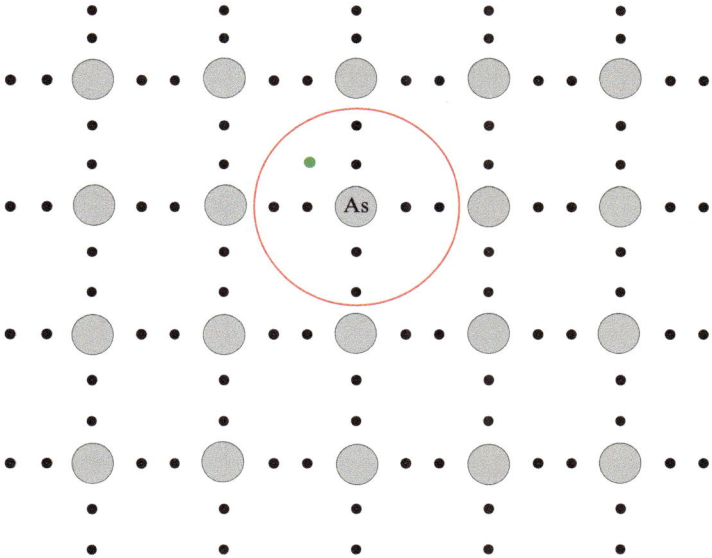

Fig. 4.2 The atom within the red ring is an arsenic atom, which provides one extra electron (*green dot*). The whole substance remains electrically neutral, but the extra electron is not bound, so it can wander around

4.1.1   *n*-Type Material

Silicon that has been doped like this (*e.g.*, with arsenic) is called *n-type*. (The electron is *n*egative, remember.) Of course, the material as a whole remains neutral, because all atoms are neutral. It's just that the "homeless" electrons that originally belonged in the arsenic can now move around freely—unattached to anything.

4.1.2 *p*-Type Material

In like fashion, we can dope the silicon with a different atom (*e.g.*, with aluminum or gallium), whereas instead of getting one extra electron, we get a deficit of one electron. We call this a *hole*. The hole acts electrically like the opposite of an electron—*i.e.*, like a positive charge, and such material is thus termed *p-type*.

Again, the important thing to remember is that these "doped" materials are still neutral.

4.2   The *pn* Junction

If the two types of material—both silicon, but doped in opposite ways—are placed in contact, something interesting will happen.

The wandering electrons of the *n*-type begin to move—diffuse is a better description—across the junction towards the "holes" of the *p*-type (Fig. 4.3).

At the junction some electrons and holes will meet up and disappear—they "nullify" each other. This region of the junction develops a certain thickness (brown in Fig. 4.3) and is called the *depletion* region. Again, the whole assembly is neutral, but the pale blue *n*-type silicon (on the right in Fig. 4.3) now becomes slightly (+) near the junction because it's lost some of its electrons.

Furthermore, the grey *p*-region (on the left of Fig. 4.3) has been made relatively negative because similarly it lost some holes.

What next? Very quickly there will come a time when a hopeful wandering electron, drifting towards the junction from the right, will find itself being dragged

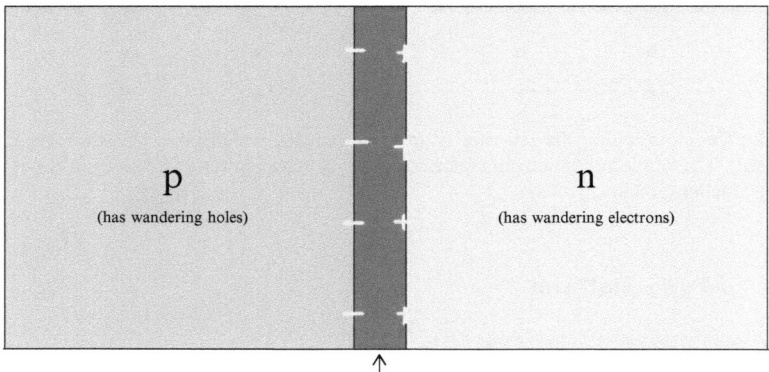

In this region, the wandering electrons
and holes nullify each other, but thus
leave the RHS relatively (+), and the LHS
relatively (−).

Fig. 4.3 The *p–n* junction. Note that the central region—the *depletion* layer or region or zone—is where the wandering electrons have met up with the "holes" and neutralized

Fig. 4.4 If we connect our external battery with the polarity as shown here, more electrons will be forced into the *left-hand side* and more (+) s into the *right-hand side*. The *brown region* (depletion zone) will then get much wider and will act as an insulator between the *grey* and the *blue*. No current will flow

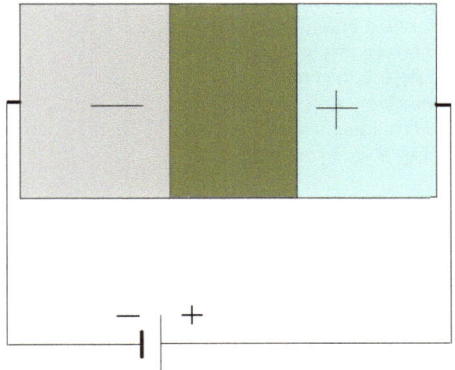

back, because, as we said, the pale blue has become slightly positive. Same with the holes on the left—that grey *p*-type has become slightly negative.

Unfortunately, this looks a bit like a battery! We say "unfortunately" because surely that's not the case?

Well, suppose we place a small *external* battery across the ends of our *p–n* junction, with its battery polarity as in Fig. 4.4. Then, more electrons will be forced into the left-hand side. The depletion layer will get wide, and we will see that negligible current will flow.

A "quick and dirty" way of seeing that no current will flow is to note that the "inherent battery" of the p–n junction is opposing the external one. Also, with the battery polarity shown the situation is often called "*reverse biased.*"

So it's tempting to say that it's similar to having two opposing batteries—and we might be tempted to predict that no current will flow, clockwise or anticlockwise. However, although this gives us the result we want, it is certainly not the explanation. In fact, with the battery polarity this way, the grey area (the *p*-type) is forced to become more (−) and correspondingly the blue area more (+). There will be more nullifications in the brown zone—which cannot conduct current because there are no free charges there—and the brown zone will then get much larger, thus becoming quite an effective insulator. No current flows.

If we place the external battery the other way around, as in Fig. 4.5, then the "quick and dirty" and incorrect way to view it is to say that the "batteries are in series," helping each other, and current will flow. However, what's really happening is that the brown zone almost disappears, leaving electrons (and holes) free to move across. And why does the brown zone contract? Because the battery, as connected in Fig. 4.5, now tends to get rid of those junction charges of Fig. 4.3, leaving a negligible insulation zone.

As always, the black arrows in this book show the conventional mathematical current, from the (+) side of the battery to the (−), and blue arrows the electron current.

We have thus made our "one-way valve." It's called a diode because *di* means *two* and it is a junction of two materials, *n*-type and *p*-type.

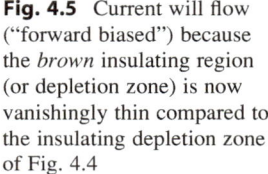

Fig. 4.5 Current will flow ("forward biased") because the *brown* insulating region (or depletion zone) is now vanishingly thin compared to the insulating depletion zone of Fig. 4.4

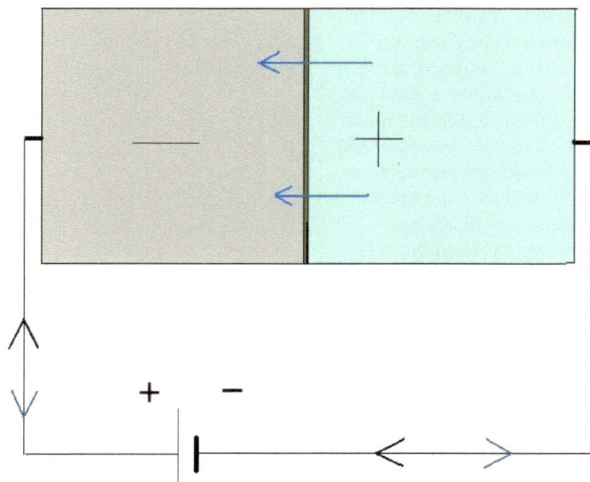

In the above we have looked only at diodes made from silicon or predominantly silicon. As mentioned, we may encounter other types of material that have been doped—*e.g.*, germanium instead of silicon. Germanium is a bit more expensive than silicon and has slightly different properties, but in the majority of experiments it doesn't really matter which type of diode we use.

4.3   Experiment: Diode Graph

In Chap. 2 we saw experimentally that a diode will only pass current one way, and now we have explained it. It didn't matter whether it was a regular diode or an LED. A more detailed kitchen experiment might be to check this, plotting a graph of voltage against current, using the universal household-type meter (*e.g.*, as shown in Fig. A.2, p. 159, or others).

We need to find a way to get different voltages, and the world does not make "1/2" V batteries. We'd like a minimum of 5 or 6 points on our graph, and one method of dividing voltages into smaller parts is shown in the following circuit, Fig. 4.6—often called a potential divider and/or a *potentiometer*:

Take a battery (9, 6, or 1.5 Volts), and directly connect some quite large (*e.g.*, 100 Ohms) series resistors across it, as in Fig. 4.6.

If the battery voltage V = 9 Volts, say, and if there are six 100 Ohm resistors, then the total current flow from the battery is 9 Volts/(100 + 100 + 100 + 100 + 100 + 100) Ohms = 9/600 = 15 mA, which won't drain the battery for quite a while. Clearly, V/6 is 1.5 V, so we can get 1.5, 3, 4.5, 6, 7.5, or 9 as possible output voltages.

Thus, to verify the graph of voltage *versus* current for a diode (which needs smaller voltages than 9 Volts) we might try a 1.5 Volt cell, such as AA, C, or D, on

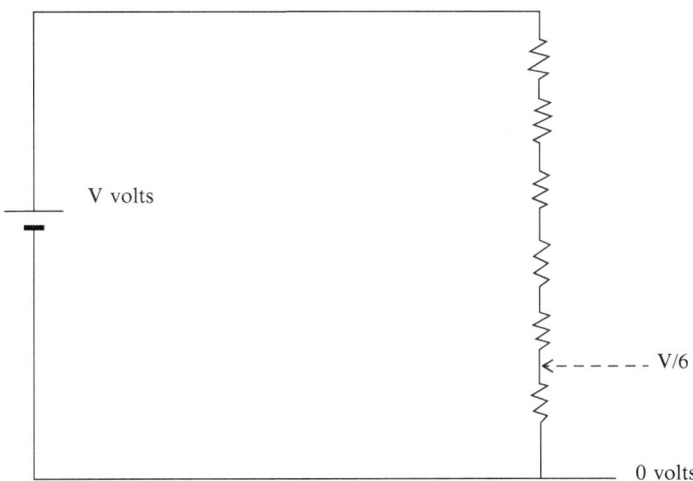

Fig. 4.6 This circuit divides a battery potential V into fractions of V (here, V/6, because there happen to be six identical—quite large value—resistors across V)

the left of Fig. 4.6, and then our system of six "dividers" would yield 0.25, 0.5, 0.75, 1, 1.25, and 1.5 Volts as possible outputs, and so on.

Because the reader may not have the patience to do this right now, another way to achieve different voltages is to use combinations of AAs, or Cs or Ds (all 1.5), and 9 Volts batteries in series.

As an example of the latter method, the following graph points were obtained for a red LED with a safety R of ~235 Ohms and with the universal household meter on its 250 mA scale [NOT the *micro*Amp scale—the meter would be damaged!]:

V (Volts)	10.5	9	6	3	1.5
I (mA)	30	25	18	5	0

These data are sketched in Fig. 4.7. (It is convenient to start at the high-voltage end to make sure that the LED is lit, and note that a typical LED really doesn't like more than about 30 mA.)

The first oddity is that, connected either way, the resistance at the very low voltages is extremely high—no current backwards or forwards, as the graph of Fig. 4.7 shows. Near the origin, in the "forward" mode (*i.e.*, in the top right-hand quadrant of the graph) no current appears until about half a Volt is applied. After that, the current increases with voltage almost linearly.

The other way round (backward mode, or bottom left quadrant of the graph) there is no current flow, as expected. More accurately, it is initially unmeasurable, being fractions of a milliamp. The voltage–current graph of Fig. 4.7 should be compared with the Ohm's law graph (Fig. 2.25 on p. 53).

Fig. 4.7 Note that when the voltage is negative there is hardly any current at all (until it is swamped and the diode destroyed). Also, when the voltage is positive, and beyond about ~0.6 V, a diode becomes almost Ohmic

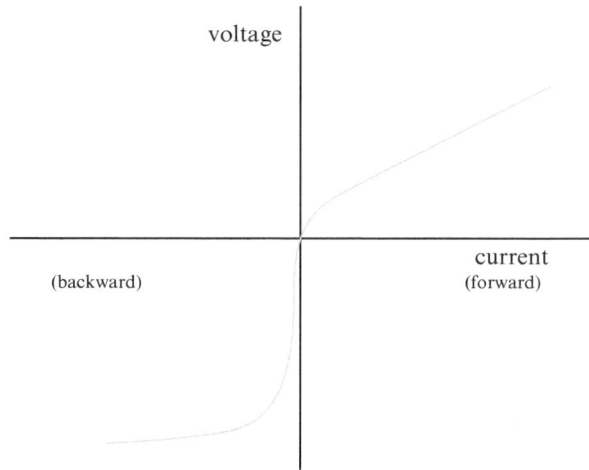

Another way to say the above, concerning the "half-a-Volt" characteristic, is to say that when the LED is functioning there is a voltage drop of about half a Volt across it. If it's a germanium type of diode this might be as small as ~1/10 of a Volt.

Generally, semiconductor diodes are unbelievably cheap—we can buy a handful of diodes for less than a dollar.

It should be reemphasized that there are many different kinds of *pn* junction or diode. There are not only the plain type of diode of Fig. 2.32 and the LEDs that we first met in Fig. 2.27 but also LEDs that can emit light of a very *narrow* band of wavelengths—essentially one wavelength—called *laser* diodes. The latter produce a thin beam of coherent monochromatic[1] light that hardly diverges. These small lasers are a great stride over the expensive and delicate glass-tubed He–Ne lasers of the 1970–1990s and are used in those pencil-sized blackboard pointers. They are also used in CD players to shine on the underside of a CD or a DVD and, by reflection, read the usual ones and zeros of the data or the music.

4.4   About Displays

We have commented on electrons and holes canceling each other out.

In general, when an electron "falls into a hole" a certain amount of energy is released, and if that energy is large enough for light to be emitted then we have our LED. Different materials for the diode will give different colors. A silicon *pn* junction for example will commonly yield red, but different colors can come from variations, such as germanium–gallium–arsenide and others.

---

[1] Monochromatic literally means "one color" and thus only one radiation frequency. Coherent means all the little wave pulses or photons are "in step" with each other or "in phase."

Fig. 4.8 Displaying the
numeral "3"

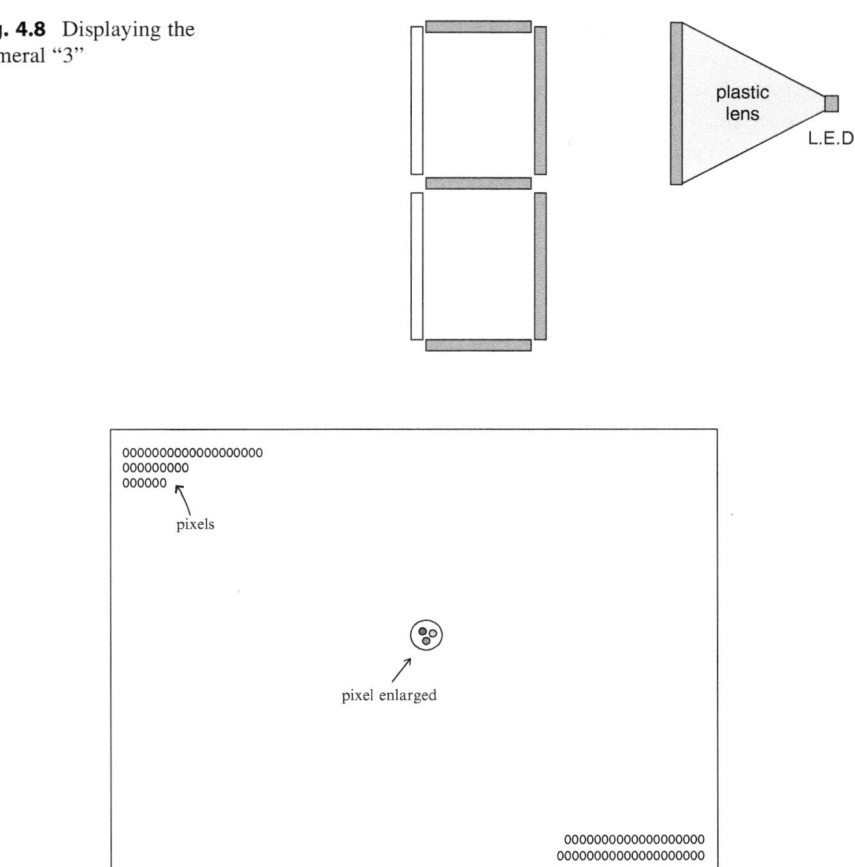

Fig. 4.9   Three tiny sub-pixels (*red, green, blue*) as one pixel in a display

Both LEDs and liquid crystal displays (**LCDs**) can, in principle, be used for displays, clock readouts, and so on (but LCDs are not sources of light, as explained below). Large billboards can even use bulbs.

Figure 4.8 shows how to form a character. Clearly, just seven segments can give any letter or number.

The light from any tiny source can be "elongated" by placing it behind a tiny lens. In Fig. 4.8 the number "3" has been illustrated. If all seven tiny sources are switched on we would get the numeral "8."

Figure 4.9 shows how three minuscule sources can in principle form a *pixel* (picture element) of light. The pixel could be made up from three differently colored LEDs or bulbs—red, green, and blue. These three are the primary colors of the trichromatic theory, developed in the 1800s by Helmholtz, as well as (separately) by Maxwell, who figures so prominently in electricity and magnetism and throughout the sciences. (Interestingly, Maxwell was one of the first to make a color photograph in the 1860s.)

Varying the intensity of any of the three primary colors will yield other colors. For example, if only red and green are on, the color would be yellow, and if red, green, and blue are all on equally, the result would be white.

4.5   Comment on LCDs

The vast majority of flat-screen TVs, in the early years of the twenty-first century at any rate, are utilizing **LCDs** rather than LEDs, despite myriad advertisements in the newspapers for "LED TVs." (The latter description can be misleading, for they are LCD displays, backlit by LEDs.)

Although the subject of LCDs is more strictly "optoelectronics," the basic ideas are the following:

1. A white backlight behind the display screen is needed; an LCD does not produce its own light.
2. This backlight goes through two optical polarizers in series.
3. Between these two polarizers is a "liquid crystal" material which can rotate the plane of polarization by an amount that depends on the applied voltage.

The net result is that if no light comes through, one of course sees a black region, and so this can be an element of the display which is black. If the polarizers are parallel then light does come through, giving a white element. The backlight often comes from compact fluorescent lamps (CFLs), but increasingly it comes from LEDs—which is the source of confusion in the newspaper advertisements.

In LCD displays, colors may be achieved in a similar way to the description on the previous page—again by utilizing three "sub-pixels." Such LCD sub-pixels would have red, green, and blue filters.

This leaves open the possibility of using ONLY LEDs, with no LCDs at all. In this case, the sub-pixels would be three little LEDs, red, green, and blue.

Recently, another type of very clear display has been developed—the **E-Ink** screen—which uses electrostatics, and we describe this in an Appendix on p. 165.

4.6   The Transistor

First, there is an elementary analogy for a transistor. Look at the "hosepipe" in the center of Fig. 4.10. Water (blue) is being emitted from some source and would flow through the hose towards a collection point, such as a bucket, past some vanes, IF the vanes were open. However, the vanes are normally closed, so the transistor is "OFF," until we do something.

(Instead of "Venetian-blind" vanes, the hosepipe can be considered squeezed shut. If we release the squeeze water can flow through.)

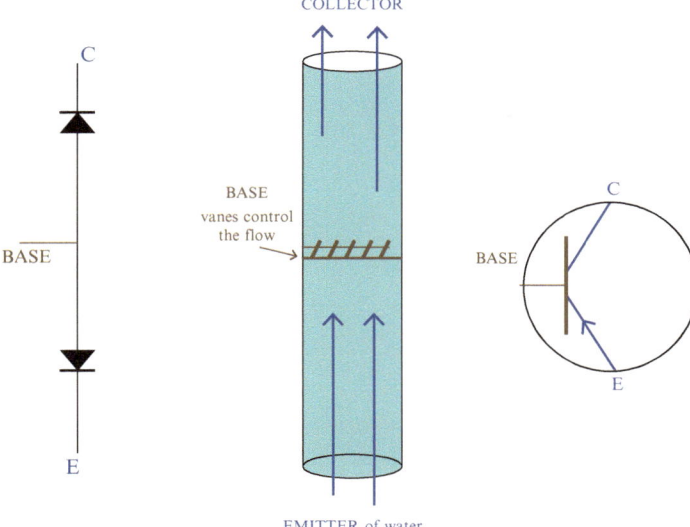

Fig. 4.10 A transistor. The *center* figure gives the hosepipe analogy for a transistor

The words used in transistor studies are EMITTER for where the water or the electrons are emitted, COLLECTOR for where the current is collected, and BASE for the squeeze (or vane controller). The transistor symbol (for a *pnp* transistor)[2] is on the right of Fig. 4.10, and on the left we have drawn two back-to-back diodes for the situation where it is squeezed shut.

As mentioned at the beginning of Chap. 4, the word *transistor* comes from a combination of *transfer* and *resistance*, because a transistor transfers, or changes, resistance to current flow as with the current of water flowing through the hose from emitter to collector.

One of the most common transistors, still used today and first made in the 1960s, is the *npn* transistor 2N2222. Transistors like these are called *bipolar* because the charge carriers are of both polarities, being both electrons as well as holes.

4.7 Experiment: Transistor as Switch

First, illogically maybe, we will do a transistor experiment and discuss it afterwards. So, back to any electronics store or the Internet to buy any cheap little common transistor.

There are three terminals on the transistor in Fig. 4.11: the E, B, and C of our analogy. (The diode only had two, unlabeled, although there was a flat side to show

---

[2] The arrow is reversed for the *npn* transistor.

Fig. 4.11 A physically big
(!) *npn* transistor beside
a dime

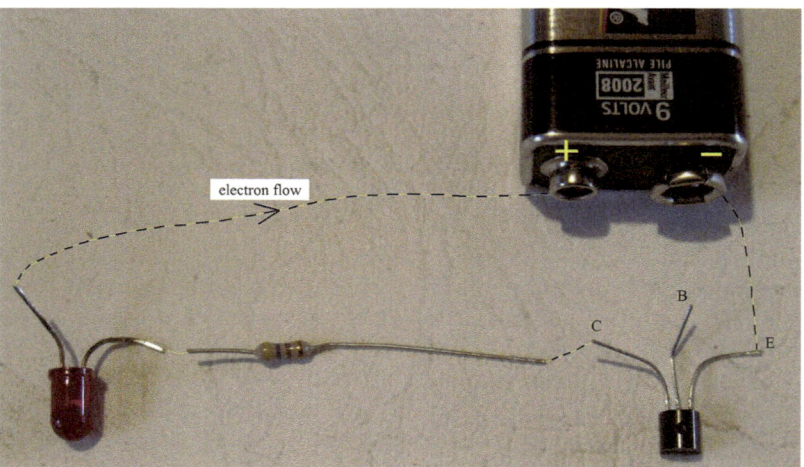

Fig. 4.12 A battery, a transistor (*bottom right*), and an LED (*bottom left*) with its necessary resistor. This experiment will illustrate the switching action of a transistor: if a small current is routed to the base *B* (which is sticking up and not yet connected) the LED will light. *Yellow lines* show the connections, rather than a clutter of wires

which way round would be forward and which way backward.) With the transistor, the E, B, and C pins are defined by the diagram on the package—so don't throw the package away!

On the kitchen table lay out the transistor, a 9 V battery, and an old LED with its all-important series resistance (try ~500 Ohms) as in Fig. 4.12.

The *symbolic* circuit diagram is shown in Fig. 4.13.

After hooking up the circuit (in Fig. 4.12 we omitted the clutter of connecting wires) we find that nothing happens. Also, recall that this is a low-voltage circuit and we can't hurt ourselves if we touch things.

Fig. 4.13 A circuit which
shows the use of a transistor
as a switch

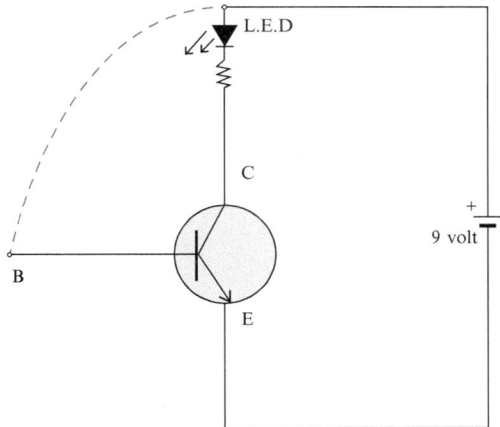

In Fig. 4.13 we have put a dotted line between the base B and the top of the
diagram, and so far no current (electron current) is flowing through the "hosepipe"
(transistor) from E to C.

If a small voltage (release of pressure on the hose) could then be applied at B
current may begin to flow, and so the LED (or small bulb) could "switch on."

So, try connecting the dotted line by means of your wetted fingers. Alternatively,
try a short string wetted with salty water. If neither of these methods light the LED
try another resistor as small as about 200 Ohms.

In any of these ways the LED should switch on.

4.7.1   What Has Happened

Electron current wants to flow up from the emitter to the collector, but doesn't, until
we do something with the all-important base B. The reason it doesn't is because the
path from E to C initially consists of two diodes back to back (see the left part of
diagram in Fig. 4.10). This is like having two check valves back to back—no water
would be able to flow.

However, if a small voltage is applied at B, the idea that the transistor is still two
back-to-back diodes changes. The depletion zone at B changes. A large depletion
zone acts like an insulator, so if something causes one of the depletion zones to
shrink, then conduction becomes possible.

The base B always turns out to be a highly sensitive controller of the current that
can flow from E to C. Thus, if we allow a connection through the wetted string, or
the finger, a small (mathematical) current can flow in at the base, and this can permit
conduction through the transistor.

A "breadboard" hookup is shown in Fig. 4.14 as an alternative to the layout in
Fig. 4.12.

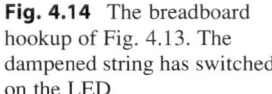

Fig. 4.14 The breadboard hookup of Fig. 4.13. The dampened string has switched on the LED

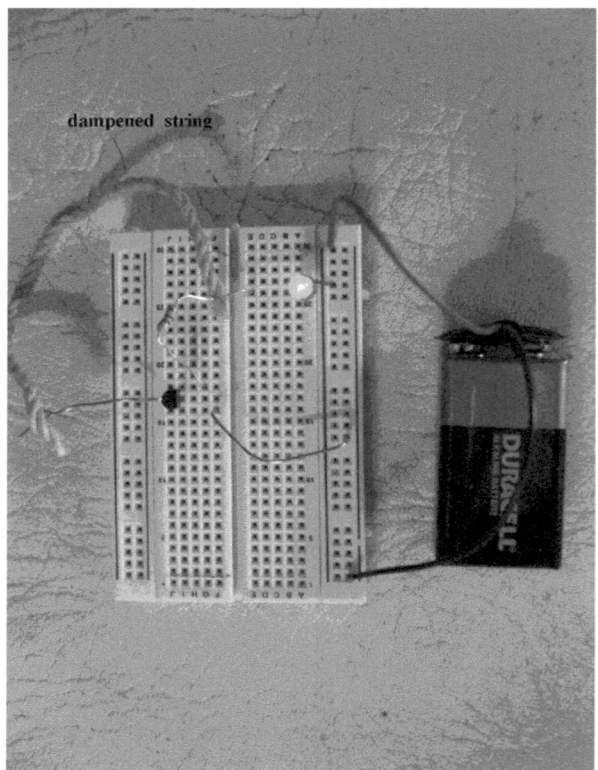

4.8   Experiment: Transistor as Amplifier

The most important thing a transistor does is amplify.

The transistor we have chosen works at a maximum power of 0.625 Watts, so it's not going to generate a massive sound. Commercial amplifiers will use a sequence of transistors to achieve high power, usually in the form of an integrated circuit (IC).

However, a small transistor will drive successfully a low-power earpiece or, as in this example, a small *"piezoelectric transducer"*[3] (but don't expect HiFi!).

Our circuit employs two of these useful devices, one as a microphone, and the other as a loudspeaker. They have the additional advantage of very high resistance, so connecting them to the transistor does not alter the voltages at crucial points in the circuit.

---

[3] A transducer is something that leads energy across (*ducere*, to lead, *trans*, across)—and in this case it uses the piezo effect that we described on p. 48. Acoustic energy on a crystal causes voltages (*i.e.*, microphone), and little voltages across the crystal cause vibrations (loudspeaker).

Fig. 4.15 A one-transistor amplifier being tested …

4.8.1   Setting Up the Circuit

In the following, for simplicity only, we will use the terms "microphone" and "loudspeaker" instead of "transducer."

1. If the loudspeaker is to oscillate symmetrically to produce a sound, its terminal connected to the collector C of the transistor must vary symmetrically in voltage, which implies that when there is <u>no input from the microphone</u> its voltage must be midway between the upper and lower possible values, *i.e.*, midway between the voltages of the battery terminals. (This state is called the *quiescent level*.)

   In our circuit we have used a 6 V battery (on the left of Fig. 4.17), so we must arrange that the quiescent level of C is +3 V relative to E (which is connected to the 0 V level of the circuit).

   To do this we must allow a constant trickle of current into the base of the transistor. Look at the BIAS resistor in Fig. 4.16. Starting with a very large variable resistance (actually 0–1 MOhm), and carefully reducing it, we found that a bias resistor of 220 kOhm would allow the "correct" base current. (By "correct" we mean giving the correct 3 V on the collector.) There is some variation in the properties of individual transistors, so exactly 3 V quiescent level may not be achieved, but this would only be significant if you were searching for maximum sound output. If you connect a voltmeter between C and E of the transistor (again with no input to the microphone) you can read the quiescent level, as in Fig. 4.17. If it's too high, reduce the bias resistance and vice versa.

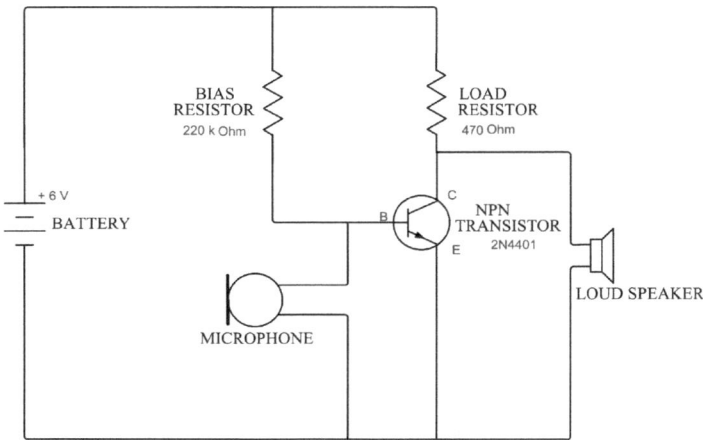

Fig. 4.16 A practical one-transistor amplifier

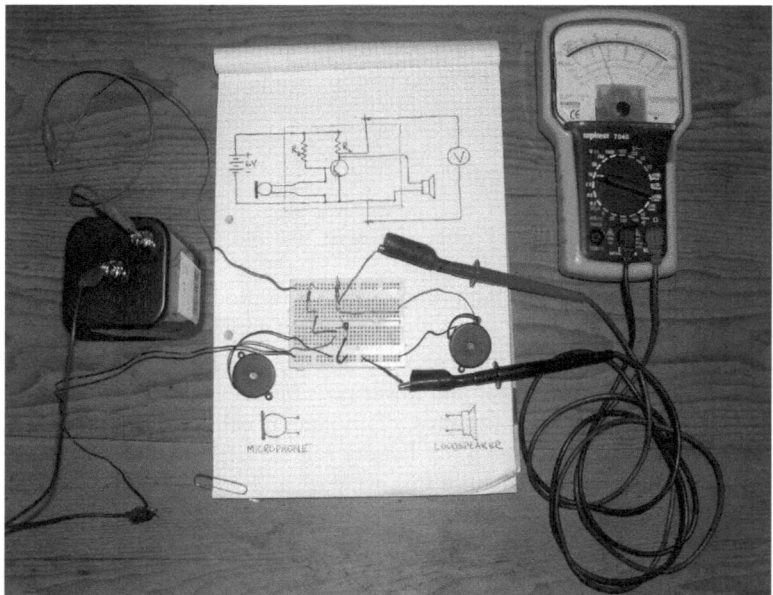

Fig. 4.17 The one-transistor amplifier circuit with breadboard connections. The voltmeter is reading 3 V

2. The load resistor controls the maximum current that the transistor can be asked to carry and so protects it from overload. When the transistor is passing its maximum current we may say that its resistance is approximately zero. We are going to choose a large load resistor—say 470 Ohms—and then from Ohm's

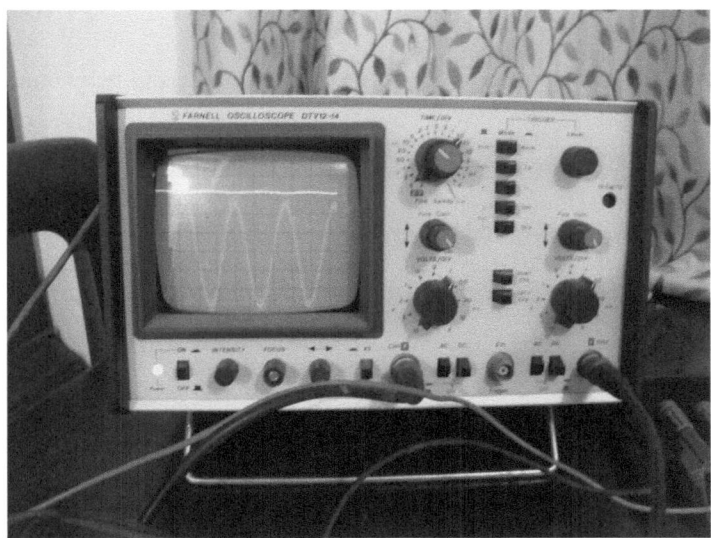

Fig. 4.18 An oscilloscope shows the amplification or *gain*, which here is about 50. The input was from a signal generator, and although these things are not generally found in the kitchen, this oscilloscope photograph is put in for interest, especially if the home experimenter has not been convinced by the intensity of sound from the crackle in his/her particular circuit

law, $I = V/R = 6/470 = 0.013$ A, which is indeed a safe current for our transistor. (A safe value for current I_{EC} should also be given on the packet.)

3. The microphone (refer back to Fig. 4.16) will also contribute current into the base. Since sound waves oscillate the diaphragm of the microphone forward and backward, this current can alternately add to or subtract from the biasing current, and this causes the collector voltage to oscillate around the +3 V level.

4. Consequently the loudspeaker emits a sound that is say 50 times louder than that exciting the microphone.

5. Figure 4.17 shows our experimental setup. The voltmeter is connected between C and E and is showing a quiescent level of +3 V. If you tap the microphone gently with a pencil, the loudspeaker emits a very satisfactory CLACK.

A short video of this experiment can be found at http://www.springerimages.com/videos/978-3-319-05304-2. (Search for the video "Making an amplifier").

As the original sound and its amplified version are sourced so close to each other, you may remain unconvinced of the amplifying effect. In this case, you can extend the loudspeaker leads by a yard or so and puff at the microphone. You will then hear a healthy PUFF from the more distant loudspeaker.

We conclude this little introduction to an amplifier by showing—yes, away from the kitchen—our input and output voltage results on an oscilloscope, which one of the authors has the good fortune to possess (Fig. 4.18).

4.9   An "Absolute" Electroscope

In Chap. 1 we made an electroscope using a kitchen jar and two aluminum foil leaves.

The following describes another electroscope—an extremely sensitive "absolute" electroscope. We call it "absolute" because whereas with the old electroscope we couldn't tell (+) from (−), this device can tell us beyond any shadow of doubt!

There are many kinds of transistors, and here we use what is called a *field effect transistor* (FET), the MPF102, which, as you can see below, has a slightly different symbol. It is interesting to note that patents concerning FETs go back to the 1920s and 1930s.

The schematic is shown in Fig. 4.19, and it is essentially the schematic for the transistor-as-a-switch on p. 135, except that the damp string has been removed.

It will be easier to construct our circuit if we move the LED across to the right, next to the battery as in Fig. 4.20. This way we can directly solder the (−) of the LED to the (−) of the battery, as the photograph (Fig. 4.21) of the completed device shows.

Nothing has changed electrically from this modification, because we know that the current will still flow around the circuit, with no turnoffs, and will thus light the LED.

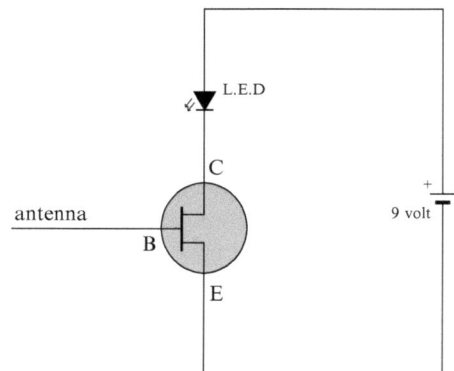

Fig. 4.19 Circuit diagram for the absolute electroscope. The reader should compare this with the circuit of Fig. 4.13

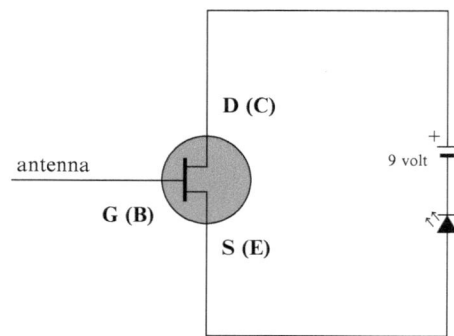

Fig. 4.20 The same circuit as in Fig. 4.19

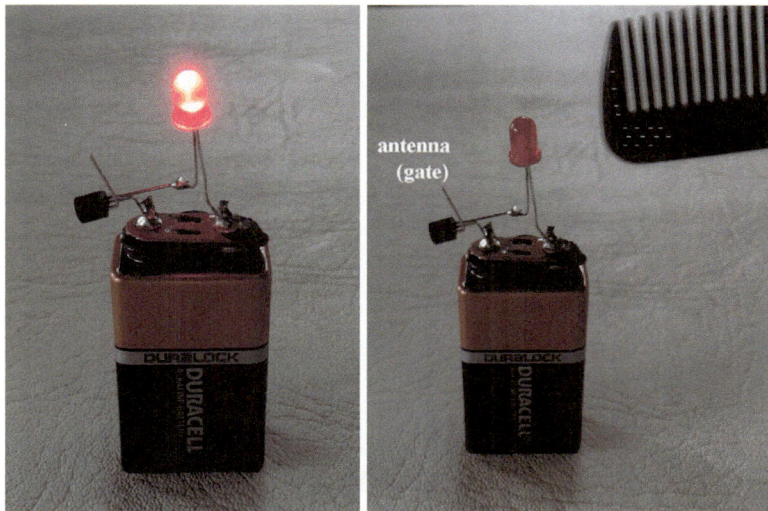

Fig. 4.21 Shows two views of the "absolute" electroscope. *Left* is normal; *right* is when a ($-$) charge is brought near

Both the transistor we used before and this one have three terminals. What we have been calling base, emitter, and collector are now, with the FET, specially named *GATE*, *SOURCE*, and *DRAIN*, respectively. This seems annoying, but no extreme harm will be done here if the beginning reader is happier with the old terms.

Now the difference between our old "transistor-as-a-switch" circuit and this new circuit is that there is nothing to supply current to base or gate B. Instead, the gate will act as an antenna. It is going to "feel" electric fields.

The circuit is trivial to construct. Following Fig. 4.20 a standard "flat" 9 V battery connector (with its plastic cover snipped off with small scissors) has its negative plug soldered to the negative side (flat side) of the LED, as in the photo of Fig. 4.21, and the positive side of the battery soldered to the central lead (i.e., the collector or the drain) of the FET. The remaining side of the FET (*i.e.*, emitter or source) is soldered to the remaining side of the LED. The end wire, the gate or the antenna, may be bent upward if you want to make it look like an antenna.

CAUTION: For those who have never soldered transistors and/or LEDs it is important not to let the hot soldering iron get too close for too long and to hold the wire being soldered with pliers or tweezers so as to conduct the heat away.

This setup is also beautifully described, on the Internet, by Wm J. Beaty, Research Engineer, University of Washington, Seattle, to whom we wish to give credit.

For those who prefer not to solder at all, the breadboard setup is equally simple and is left as an exercise because it is so similar to the breadboard setup of the transistor-as-a-switch.

4.9.1   Using the Absolute Electroscope

The left photograph in Fig. 4.21 shows the FET (black, on the extreme left) lighting the LED, and this is its normal state—*i.e.*, when there is no charge nearby. The right photograph shows the approach of a charged comb, which has caused the LED to switch off.

We are then forced to conclude that the nearby comb, rubbed through the hair or on a cloth just as in Chap. 1, has influenced our circuit. Actually, if the comb is negative the LED switches off; if positive, it glows brighter—brighter than in the left photograph. But what's going on?

4.9.2   What Is Happening

The key idea is that FETs control the current flowing between source and drain (and consequently through the LED) via the electric field felt by the gate—NOT by any current flowing into or out of the gate, as with the first transistor we used, the so-called bipolar transistor.

From the circuit of Fig. 4.19 there can't be any current flowing into or out of the gate, for there is no wire connected there! There is only the manufacturer's lead, which we bent upwards. If this antenna "feels" a field and/or a potential—from a nearby comb, say, then depending on whether the field comes from a (+) or a (−) this is what controls the light we get from the LED.

In summary, and repeating ourselves, NO light means (−), much BRIGHTER means (+), and medium brightness means zero field.

Thus we have come full circle. We are able, thanks to the FET, to tell simply and cheaply what the polarities of those electrostatic charges on balloons and combs actually were.

Note that the sensitivity of our electroscope could be increased by lengthening the antenna, but this is not necessary. Also, do NOT touch a charge to the gate/antenna, for this could easily destroy the FET.

Interesting little party tricks could be performed with our absolute electroscope. Try scuffing your shoes on the floor while holding the electroscope. Try jumping up and down with it; you will find that you are charged differently on or off the floor.

4.10   Connection Between Fields and Potentials

In the previous experiment the reader may be left wondering why we have said "electric field" OR "potential"—and why these two apparently different things are directly associated.

Referring to Chap. 1 (p. 18) where we sketched the electric field $\mathbf{E}$ from a point charge, as well as between the plates of a capacitor, we may now add lines of *equal potential*, called equipotentials (the dashed lines) to the sketch, as in Fig. 4.22. And in passing, note that the dashed lines are always perpendicular to the field lines.

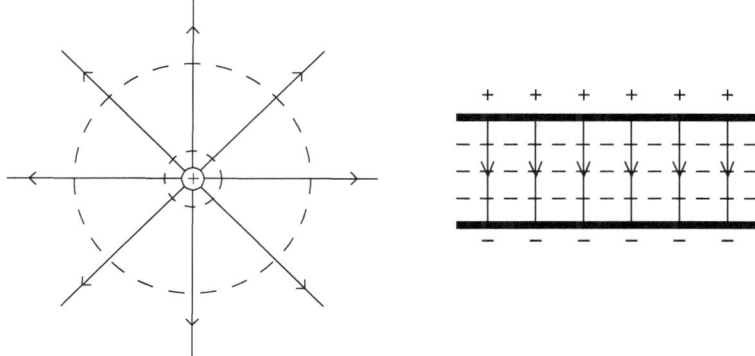

Fig. 4.22 Along any *dashed line* the voltage is constant. Such lines are called "equipotentials," and they are always perpendicular to the **E** field

But now, how is it possible to say that there is a voltage, or a potential, in space? Don't we always think of a voltage as being on the terminal of a battery or on some metal object?

Consider the capacitor first. Imagine a 9 V battery across it, with one plate being at 0 V and the other plate at +9 V.

Now imagine a tiny free pith ball, or test charge, carrying a (+) charge, very close to the (+) plate. (We are neglecting gravity completely in this discussion.) It will be repelled from that (+) plate, and simultaneously attracted to the (−) plate, and will have quite a bit of potential energy—more than say if it was starting from somewhere in between.

This is analogous to raising a ball in the air—the higher you go the greater its potential energy or ability to do work. The higher the waterfall the more energy or work (driving a mill wheel, say) we can get.

If the pith ball were at the halfway mark, it would have half the electrostatic energy than at the beginning—and it has already done some work. In fact, the pith ball (q), when adjacent to the (+) plate, has potential energy (q Coulombs × 9 Volts) as we said on p. 40, but halfway down it has already moved along the field line towards the (−) and can now only have half this amount, *i.e.*, q Coulombs × 4.5 Volts.

So we say that there are indeed lines of "voltage" in space, analogous to contour lines in geography.

In the case of the (+) central charge (left side of Fig. 4.22), the closer the little (+) pith ball, or test charge, is to the central (+) charge, the more potential energy it has, because the field lines are closely concentrated there. The **E** field is strong, and the repulsive force is great.

Further away, the field lines are less concentrated; the **E** field is weaker, and so the repulsive force is less. The ability of the little pith ball to do useful work is thus less. Since the pith ball's charge is still a constant "q," the potential energy, and therefore the potential "V," in space, must be getting smaller the further we are away from the central point charge.

4.11   Experiment: Charging and Discharging Capacitors

If we momentarily connect a battery across the plates of a capacitor, it becomes charged until the voltage across its plates becomes that of the battery. This much we know. But let's monitor the event with a voltmeter—either analog or digital.

The analog meter gives a good idea of what's going on, but it alters the characteristics of the circuit in which it is connected, and for strict accuracy an allowance would need to be made for this. (Further discussion was in the multimeter section in Chap. 2, p. 70)

If we use a digital meter, it has almost no effect on the behavior of the circuit, so its readings are accurate as they stand, but interpreting a changing digital display is less intuitive.

It's possible to connect the meter probes to the plates, and then connect the battery terminals, while watching the scale on the meter, but you may well feel that you need three hands or possibly two people. It's much easier to use the "breadboard" (prototyping board) used in other of our experiments. The breadboard gives you the equivalent of many more hands!

In Fig. 4.23 we are using an *electrolytic* capacitor which packs a huge capacity into a small volume. These tend to look like a miniature beer can and have their terminals marked (+) and (−). They must only be charged according to the polarity indicated, although other types of capacitor can be connected either way.

We have connected the (−) of the 9 V battery to a row of holes along the bottom edge of the breadboard and the (+) (red) to a row along the top. Thus we can easily make a connection to 0 V or +9 V anywhere on the board using these rows which

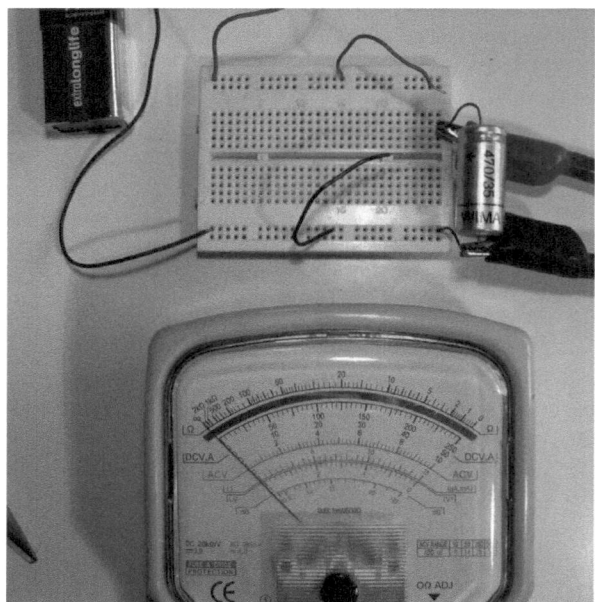

Fig. 4.23 Breadboard setup of circuit to begin charging the shiny electrolytic capacitor on the *right*

are called "busbars" (Latin, *omnibus*, "for all"). The $(-)$ "leg" of the capacitor is plugged in to the $(-)$ busbar. A "flying lead," color coded red, is plugged into the $(+)$ busbar. To make it easy to discharge the capacitor completely—we're going to do this numerous times—there is also a black "flying lead" plugged into the $(-)$ busbar; touching this to the $(+)$ plate discharges the capacitor instantly.

For this first trial we used a capacitor of 250 μF; small capacitors and small resistors are supplied in a series of "preferred values" that are convenient to the manufacturer, not necessarily the user. We used an analog meter.

When we touch the $(+)$ flying lead to the leg of the capacitor connected to the $(+)$ probe of the meter, the pointer zooms up to 9 V as you would expect. But even if you don't discharge the capacitor, you can see that its voltage immediately starts to drop. This discharge seems to be a slow process, and as you watch, it slows down more and more; after 2 min it's still showing about 2 V.

How can it discharge at all? There must be some conductive route from the $(+)$ plate back down to 0 V, and the only such route is through the voltmeter itself, as we saw in the multimeter example in Chap. 2.

These observations tell us at least two things:

1. That the voltmeter is playing a significant part in the experiment: A digital multimeter with much higher internal resistance would be better in this respect; but interpreting changing readings would be difficult. Analog devices give you an instant impression; and that's why we like analog clock faces and speedometers.

2. That a high-value resistance slows down discharge, and presumably introducing a similar resistance in the charging circuit will slow that down too. So let's try putting a 100 kOhm resistor in the charging circuit, labelled R in Fig. 4.24:

Now, start charging the capacitor by plugging the red flying lead into the far leg of the resistor as shown. **WARNING**: Never have both flying leads connected to the resistor at once; this will kill the battery!

You will see the capacitor voltage rise, quite rapidly at first, but with the rate of increase slowing down. It won't get to 9 V because of the discharging effect of the voltmeter, which is working against the charging current.

Now disconnect the red flying lead, and connect the black one. The capacitor voltage will fall, rapidly at first, but slowing down.

If you like drawing graphs, you can sit down with a watch on the table next to the voltmeter, write down the voltage (charging) at say 10-s intervals, and plot a graph of voltage against time and another from the fully charged voltage, say 9 V, discharging.

Here a digital meter is actually more convenient and accurate, but the analog meter is still acceptable.

With more sophisticated instrumentation we would find that the two graphs are in fact mirror images and are examples of *exponential buildup* and *exponential decay*. (Runaway exponential growth, as in the increase of virus in an infection, is another, mathematically related situation.)

Your two graphs (on the same axes) will come out as in Fig. 4.25:

Why does the system behave in this strange way?

It has a close resemblance to the coffee water demonstration in Chap. 2.

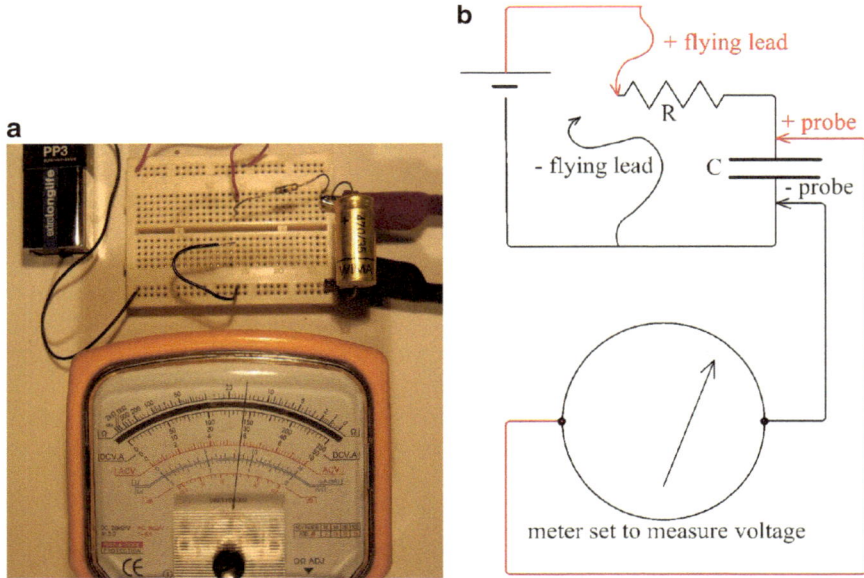

Fig. 4.24 A 100 kOhm resistor slows the charging of the capacitor

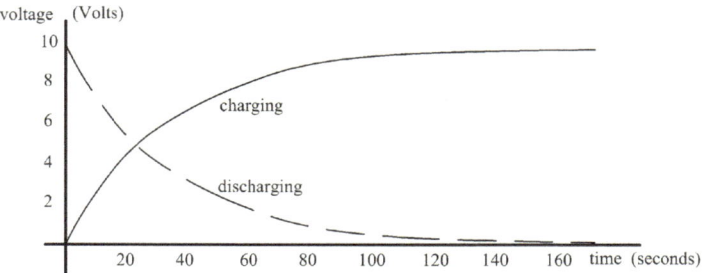

Fig. 4.25 Charge and discharge graphs for a capacitor

When the capacitor commences to charge, the (+) 9 V of the battery drives current through the resistor to the (+) plate of the capacitor, which is initially uncharged, *i.e.*, at 0 V. So the potential difference is + 9 V, and we could calculate the resulting current. We won't bother, because our rather unsophisticated voltmeter is affecting the experiment, and we would have to make a correction for that; it's the pattern of behavior that's important. But as more and more charge accumulates on the capacitor's (+) plate, it starts to build up a potential (or voltage) on that plate. For instance, in our experiment, after 30 s the capacitor was charged to a level of

5.7 Volts, and so there was a voltage of only $(9-5.7) = 3.3$ V between the battery (+) terminal and the capacitor (+) plate to drive the current which will continue the charging process. So charging will have slowed down dramatically.

When you observe the <u>dis</u>charge behavior, of course the capacitor (+) plate is initially at 9 V and drives a strong current down through the resistor to 0 V. But in a comparable way, after 30 s the capacitor will have lost a considerable amount of charge, and its voltage will have dropped to around 4.1 V, which will now be driving a much weaker discharging current.

The coffee water demonstration in Chap. 2 is an exact analog of the discharging experiment above. In the left-hand photo of Fig. 2.10 (p. 39) the bottle (capacitor) is charged to a peak level above the outlet (9 V), and the discharging current is visibly strong; in the right-hand photo the water level has fallen to about a third of the original height above the outlet, and the discharging current is much weaker, so the water potential (voltage) is dropping much more slowly.

<u>Further (optional) experiments:</u> The values of R and C suggested here are readily obtainable, but any comparable values will be suitable.

1. Replace the resistor with 10 kOhms and again with 1 MOhm. Note how the time scale of the experiment changes.
2. Replace the capacitor with 100 µF and again with 470 µF. Note yet again how the time scale of the experiment changes.
3. Revert to $R = 100$ kOhms and $C = 250$ µF, and now measure the charging <u>current</u> instead of voltage. Don't forget to rearrange the meter probes so that they are in series with the capacitor, as shown in Fig. 4.26 above.

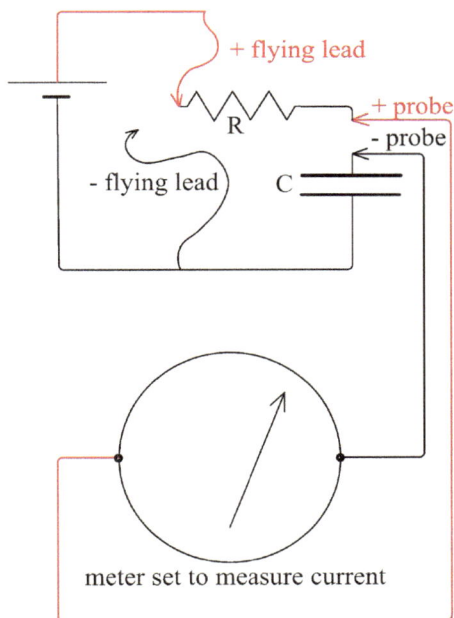

Fig. 4.26 Rearrange the probes to be in series with the capacitor

You need to switch the multimeter to <u>current</u> readings of course; the most sensitive, 50 µA, should be suitable. Now, repeat, but this time measuring the <u>dis</u>charging current.

In each case, the current starts strongly and then fades away, and in each case the voltage driving the current starts high and fades away. This is an even closer match to the coffee water demonstration in Fig. 2.10. There are many permutations of these trials, but we've done enough now to get the picture.

These exponential curves have several interesting properties, for which the math is quite sophisticated, so we won't go there! However, we can say that their analysis requires the use of a "magic number" always called e or officially Euler's number, after the great Swiss mathematician, **Leonhard Euler** (1707–1783).

e, which comes directly from a consideration of the growth of money in compound interest, has fundamental significance to processes of growth and decay in the same way as π has significance to the geometry of circles and spheres.

A consequence of this for capacitor charging and discharging is that the graph is absolutely consistent for any particular combination of R and C, regardless of the battery voltage, and exhibits a *time constant* T, which is calculated by multiplying R in Ohms by C in Farads. Most capacitors are small fractions of a Farad, which we must be careful to get right. So for our first trial, R = 100 000 Ohms, C = 250 µF, which is 0.000 250 Farads, and so T = (100 000 × 0.000 25) = 25 s.

If we tried R = 10 000 Ohms, but still kept C = 250 µF, T would be 2.5 s.

The practical significance of this **time constant** is that, when charging, the voltage takes time T to reach about 2/3 of its final level, and when discharging it takes time T to fall about 2/3 of the way to complete discharge. This is absolutely repeatable, even if the battery is losing voltage.

In electronics there are millions of useful timing circuits, which work by charging a capacitor when you start the device and detecting when the voltage has dropped to a certain level (which doesn't have to be the 2/3 level). When the specified level is reached, the buzzer goes off. In an electronic kitchen timer, the setting knob controls a variable resistor which provides the range of timings you need. The very common "555" timer introduced next will describe all the clever circuitry to convert the RC charging process into a practical device which is easy to build and reliable in performance.

There are other examples of exponential increase and decrease. If you plant a young apple tree, you will probably see no apples in the first year. But in succeeding years you will get perhaps four or five apples the first fruitful year and then an increasing harvest year by year following the buildup pattern, which eventually levels off to peak production, determined by genetics, size of mature tree, and climate. Of course, large fluctuations in climate from year to year will distort the graph!

After some years, the tree will experience a decline in vigor and will exhibit an exponential decrease. This process may well go on for decades, and after even 50 years it will bear a few apples every year.

Radioactive decay is a classic example of exponential decrease. The level of activity of a source, as measured for instance by a Geiger counter, decreases as the amount of remaining undecayed material decreases. Again, compare the coffee

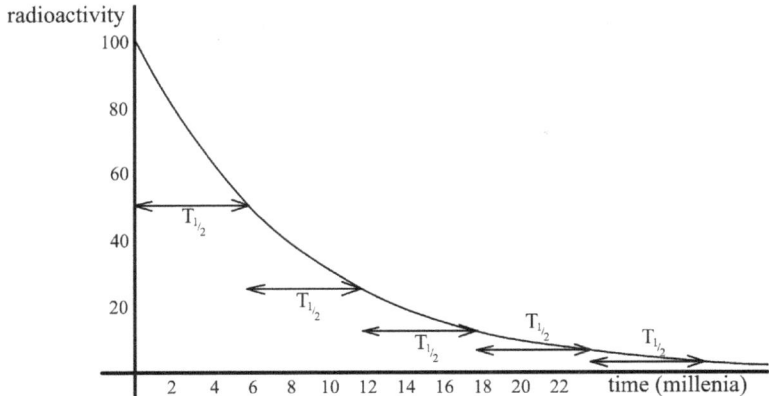

Fig. 4.27 Illustrating half-life for $^{14}_{6}C$

water demonstration in Chap. 2. In this case, unlike either the coffee water experiment or the capacitor experiment, there are no factors we can control, like R and C, to alter the time scale of the graph, but we can describe it in a readily understandable way by the quantity *radioactive half-life*.

This half-life is the time in which the activity declines by a half, symbol $\mathbf{T_{1/2}}$, and it repeats throughout the decay process, as shown here (Fig. 4.27):

Two kinds (isotopes) of carbon occur in the earth's atmosphere, in a more or less fixed proportion. The predominant isotope, $^{12}_{6}C$, often called C-12, is <u>not</u> radioactive. The radioactive isotope $^{14}_{6}C$, commonly called C-14, has a half-life of 5730 years. All growing plants absorb this mixture during their life. At death the absorption of carbon (as carbon dioxide) ceases, and from that point on the $^{14}_{6}C$ decays, with exponential decrease, while the $^{12}_{6}C$ content remains fixed. In an archaeological specimen such as an ancient timber stake, the ratio of $^{14}_{6}C$ to $^{12}_{6}C$ will have halved in 5730 years, so by measuring today's proportion we can calculate the age since the death of the specimen.

There have been significant fluctuations in this latter ratio in historical time, and the current ratio of objects of <u>known</u> age is used to correct for this.

The method can also be used for animal remains because the food chain for all animals will include herbivores at some point.

The old trick question about the frog wanting to get to a pond and being allowed only to hop such that each succeeding hop is again halfway there—and how many hops does it take to get there?—is another, and not trivial, example of exponential decrease. Of course, it <u>never</u> quite gets there!

4.12 Integrated Circuits: The 555 Timer Chip

The idea of making an electronic circuit in which all the components are combined in one object was floating around in the 1950s. The breakthrough happened when experimenters realized that a piece of silicon could be treated by local "doping" and

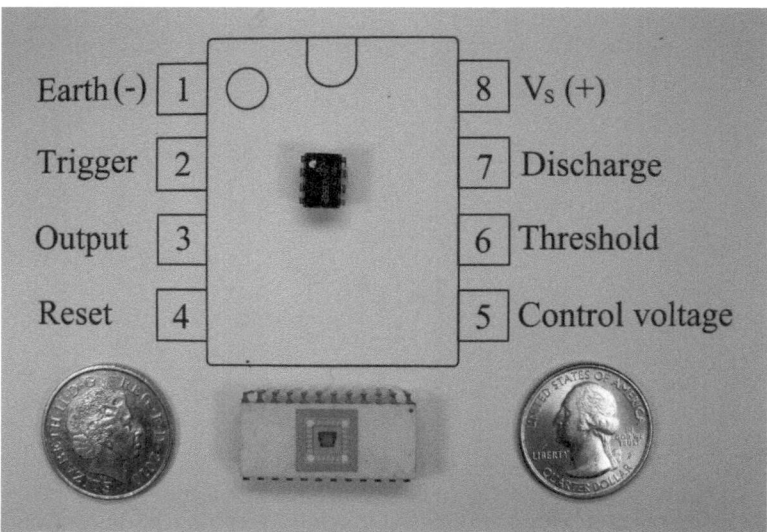

Fig. 4.28 The IC chip has a *white dot* at *top left*. For size comparison there is a UK 10p coin on the *left* and a US quarter on the *right*. A more complex DIP-24 integrated circuit is also shown

other processes so that separate areas could be made to be insulators, conductors, semiconductors, or dielectrics. So you could build up a complex of transistors, diodes, connectors, capacitors, and resistors all within the one piece. The first working IC was produced by Fairchild Semiconductor in 1960.

ICs have many advantages over the equivalent discrete circuits, *i.e.*, those assembled from separate components. They are more compact, less expensive, and more reliable and use less energy. Also, because the links between the internal components are short, digital circuits are faster working.

Furthermore, by the time an IC is put on the market, it has been thoroughly debugged, and circuits built with it will require less development and testing.

The technology is too complex to describe further here, but excellent information is available on the Internet. We can say however that in almost all cases thin slices or *wafers* are cut from a large single crystal of silicon, just like slicing a carrot, and hundreds of identical ICs are replicated on the wafer by *photolithography*, a kind of printing process. The wafer is then sawn up into individual ICs, which might be around 0.1 inch square, depending on the complexity of that particular circuit. When you buy an IC, almost all the space in the device is taken up by the connections to the outside world, which are much bigger than the chip itself. Most ICs are packaged in a little brown plastic block like a small square of chocolate (Fig. 4.28), with two parallel rows of terminals, called legs, which conveniently plug into the "breadboard" used in this book. This style is called dual in line (DIL).

The 555 timer chip was designed in 1971 by the Swiss-born inventor and electronics engineer, the late **Hans R. Camenzind** (1934–2012), working as a consultant to Signetics Corporation. He was the author of many technical papers and books as well as of a general interest book on the history of electronics.

The chip is still currently in large-scale production by prestigious companies like Texas Instruments and countless small factories in Southeast Asia, and nobody knows how many have been made, although Camenzind suggested in an interview in 2003 that production then was about one billion per year!

The 555 is a relatively simple circuit with about 25 transistors and various other components, and there are many updated versions, all using the same external connections. Because of its simplicity it only needs eight connection legs, a form called DIP-8. The reason for its continuing popularity is its versatility; it can be used as a one-shot timer, *e.g.*, to keep an indicator light on for say 30 s; to give a beep after 5 min, *e.g.*, for an egg timer; to flash a light, *e.g.*, ON for 5 s and OFF for 2 s, as in an automobile direction indicator or cyclist's rear light; to generate a beep; and so on. And that's only a few examples.

Here is a "pin-out" diagram to show the connections. The legs or *pins* are numbered 1–8, and a little dimple is moulded into the case to identify pin 1. In the photo of Fig. 4.28 this dimple has been highlighted with a white dot.

We have taken the opportunity to include in the figure a much more complex IC, in configuration DIP-24, which has been produced as a demonstration without the plastic casing, so that you can see the chip and its connections to the pins.

The function of each pin in the 555 IC is as follows:

Pin 1.   Ground or zero Volts

Pin 2.   Trigger: Normally at the supply or the battery voltage V_S, but starts the timing process if it is connected to 0 V

Pin 3.   Output: When the timer is switched on, this is connected internally to the battery (+) to activate the light, beeper, relay, etc., which is controlled by the device

Pin 4.   Reset: Puts the timer back to its starting condition, ready to repeat the timing process when triggered

Pin 5.   Control voltage: Alters the whole time scale of the process. In many simple applications this is not needed and the pin may not be connected!

Pin 6.   Threshold: The voltage on this pin is normally around 0 V, but if it is raised to a certain level (the threshold level, of course), it causes the output pin to fall to zero Volt. And the threshold level is by design 2/3 V_S. Does that ring a bell?

Pin 7.   Discharge: This pin is normally connected internally to 0 V, and so it stops the capacitor getting charged but is disconnected internally when the trigger (pin 2) is activated

Pin 8.   It is connected to the supply or the battery voltage + V_S and supplies the energy to work the IC and also to drive whatever device is activated by the output (pin 3)

The 555 timer is remarkably flexible as to the supply voltage, which is usually in the range 3–15 V. Within this range its output timing does not vary.

555 timer ICs are readily available (see the components Appendix) at a cost of about 40 cents each.

There are also ICs with more pins, 14 or 16, which contain two or four 555s in the one package. (See below.)

4.13   Experiment: A Metronome Circuit

The following example of a metronome circuit produces the regular click which helps to keep to the beat when playing a musical instrument. (If you wanted to have a different click for the first beat of each bar, you would use a 556 chip containing two 555s.)

The IC is here set to work in an alternating or *astable multivibrator* mode. The output pin continuously alternates between V_S (which we call "high") and zero. It drives our handy piezoelectric transducer, which clicks every time the output flips high.

Here is the circuit diagram for the metronome (Fig. 4.29). Note that in the diagram, the pins are placed so as to achieve maximum clarity, not in their actual locations. The photo showing the physical build of the circuit (Fig. 4.30) on our breadboard is electronically identical, but less easy to follow.

Components List:

Essential items	Optional items
Breadboard	C_2, capacitor electrolytic 33 µF (if you use a different transducer)
Battery 9V	C_3, capacitor 0.01 µF (if you get interference from nearby electrical equipment)
R_V, variable resistor, 0–50 kOhm	
R, resistor 1 kOhm	
C_1, capacitor electrolytic 33 µF connecting wires (short lengths)	

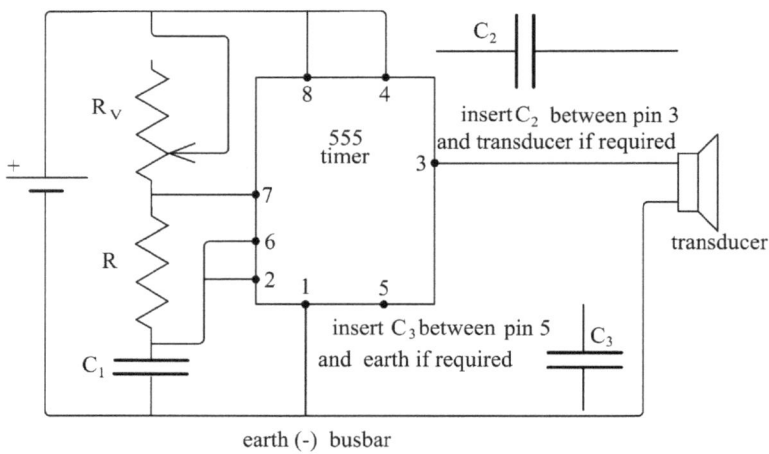

Fig. 4.29 Circuit for the metronome

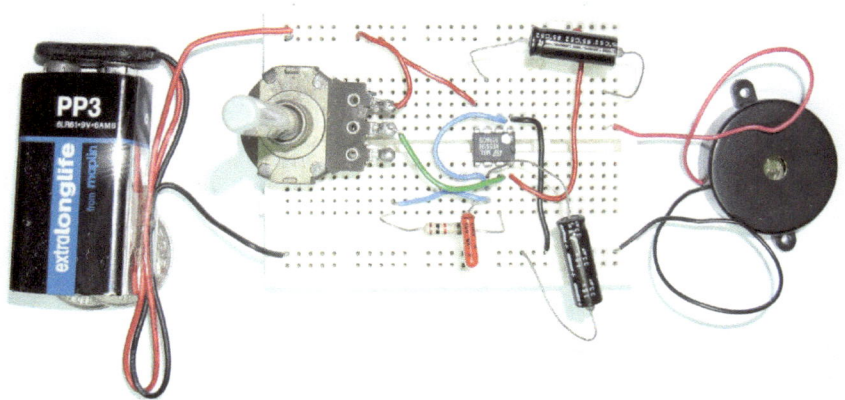

Fig. 4.30 The actual build. (The two capacitors on the *right* are optional)

Because you can adjust the click frequency by means of the variable resistor, component values are not critical; just use the nearest value you have to hand.

Here is how it works (if you just want the satisfaction of hearing the metronome's clicks, you can skip this bit):

1. When you switch on (*i.e.*, connect the battery), pin 6 (threshold) is at 0 V, and C_1 is not being charged. But it is connected to pin 2 (trigger), which is immediately lowered to 0 V, and starts the timing sequence.
2. Pin 3 (output) flips high, the transducer clicks, and pin 6 is disconnected from 0 V.
3. Capacitor C_1 starts to charge via R_V and R.
4. When pin 6 (threshold) reaches 0.7 V_S, and the capacitor is approaching fully charged, pin 7 (discharge) is reconnected internally to 0 V, and pin 3 (output) flops back to 0V. The transducer is no longer energized.
5. So C_1 discharges through R, quite quickly because R is only 1 kOhm, and pin 6 flops back to 0 V.
6. But when it does this, it makes pin 2 (trigger) 0V again which resets off the whole cycle. So pin 3 (output) flips high, the transducer clicks, and the whole cycle starts again!

The time constant T for charging the capacitor will here be $(R_V + R) \times C_1$, and there are arcane mathematical reasons—which we don't consider in the kitchen—for the metronome click time to be about 10 % greater than this. So, by altering the setting of the variable resistor we can set the click time of the metronome to essentially any value we choose.

A short video of this metronome in action can be seen at http://www.springerimages.com/videos/978-3-319-05304-2. (Search for the video "Making a metronome").

Appendix A: Resistor Color Codes

The resistor in Fig. A.1 has a value of 52 Ohms.

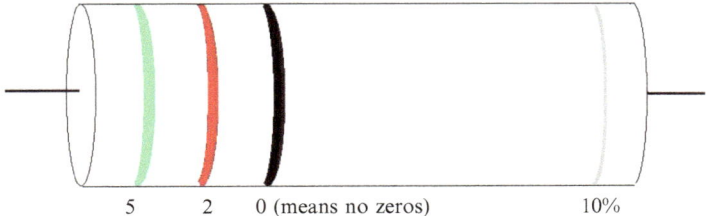

| 5 | 2 | 0 (means no zeros) | 10% |

Fig. A.1 A 52 Ohm resistor

Look at any resistor so that all the colors are clustered on the left as above. The first color here gives a 5 (green), the second is a 2 (red), and the third (black) gives the number of zeros after that. Thus, **5-2-and no zeros**, or 52 Ohms. If a resistor has been made accurate to 10 % of the claimed value the last stripe will be silver, and, for 5 %, gold.

0	Black
1	Brown
2	Red
3	Orange
4	Yellow
5	Green
6	Blue
7	Violet
8	Grey
9	White

In practice, it's often easier to check resistor values on the "Ohms" scale of any multimeter.

D. Nightingale and C. Spencer, *A Kitchen Course in Electricity and Magnetism,*
DOI 10.1007/978-3-319-05305-9, © Springer International Publishing Switzerland 2015

Appendix B: Components

Components that may be useful, in approximate order of mention in the book:

General: First, do obtain a supply of crocodile/alligator clips!

Mini alligator clips (e.g., pk 10 mini-clips)	~$2.50
Screw-in flashlight bulb (e.g., pkg of two)	~$1.20
Screw-in lamp base (estimated cost)	~$1.30
Battery holder (AA)	~$0.90
Coil of hobbyist's connecting wire	~$1.00
Small breadboard (useful in Chap. 4)	~$4.00

While in a hobbyist's store, or on the Internet, you might consider

Solar cell (0.45 Volt—three in series)	~$4.95 × 3
Small packet of LEDs (assorted colors)	~$1.50
Rechargeable AA (e.g., Nickel Metal Hydride, pk of two)	~$8
Small plain diode (pk of two)	~$0.50
Germanium diode	~$1.50

Batteries: While any 1.5 V batteries will be satisfactory for our experiments, from the tiny "AAA" used in TV remotes to the fatter "C" and "D" cells, one would then need the particular battery holder of course. For simplicity we have mostly used just common AAs. Our photographs show an occasional fat 6 V battery (*e.g.,* with the electric motor—but not crucial), and we can generally make do with combinations of smaller ones.

Resistors and capacitors: One can purchase an assortment of resistors (10 Ohms–10 MegOhms) for ~$3 on eBay, and 100 capacitors (1–470 microFarad) for ~$6.

Variable resistors (sometimes called potentiometers)—rotary or slide—are typically listed at about $1 each, or ten for ~$3.50.

Magnets: Refrigerator magnets work with some of our experiments, but really strong magnets, for later, may be purchased cheaply online, with some examples

D. Nightingale and C. Spencer, *A Kitchen Course in Electricity and Magnetism*, 157
DOI 10.1007/978-3-319-05305-9, © Springer International Publishing Switzerland 2015

already mentioned on p. 119. We suggest the reader put into something like eBay *Neodymium block magnet 3/8"*. These may be fairly thin but will do perfectly well, and a pack of eight or ten would be best. Amazon.com would also be a good source.

Meters: There are many general-purpose hobbyist's meters available online, or in hardware stores, for under $25. Be sure to choose a <u>sensitive</u> DC current scale <250 microAmps (try for 100, or even 50 microAmps). We have illustrated ours in Fig. A.2.

Transducers: One can buy half a dozen little transducer discs for ~$4 on eBay or other electronics stores. (Such discs are commonly used by guitar players, to give a signal for electronic amplification.)

Transistors: A common general purpose *npn* transistor is the 2N2222.

The FET transistor we used, MPF102, is very cheap: ten were sent promptly by airmail from China—10 for $5—from: zhangsheng FuHua Road,110, Guangye Building, East Block, 30E Futian District, Shenzhen GUANGDONG, CHINA 518033.

IC chips: The 555 chip (used only at the end of the book) is available from eBay for $5, or for $1.49 (Thailand), or 25 of them for $7.95.

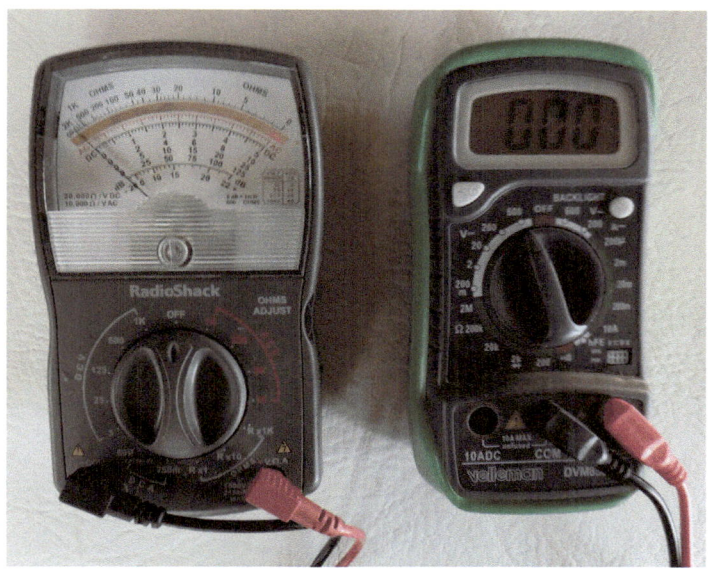

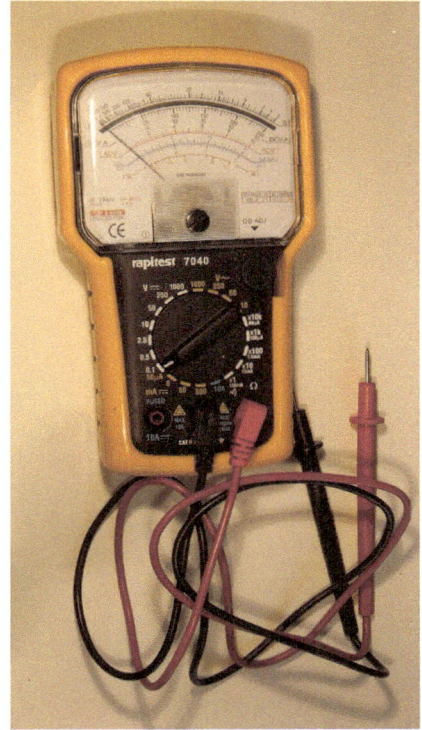

Fig. A.2 Some typical meters

Appendix C: RFID—and Bar Codes

Today, the radio waves of Heinrich Hertz are everywhere, and almost everyone carries a radio transmitter/receiver in his/her pocket. Cell phones, or mobiles, sending and receiving radio waves are not only universal, but radio waves are also used to lock and unlock car doors, and to detect and tag items in stores.

Just as our kitchen radio needed no battery, and yet was able to give tiny signals, so a "tag" can receive a radio wave and a tiny voltage may be induced—just enough to weakly *retransmit* some radio waves over short distances—revealing certain properties about itself. This kind of thing is referred to as RFID—Radio Frequency Identification. A common frequency used is 13.56 MHz (*i.e.,* a wavelength of about 60 feet) but other frequencies such as 3000 MHz (with wavelength of ~4 inches) are also used.

RFID tags, with no battery, are seen on the windshields of cars, trucks, and buses as shown in Fig. A.3.

The minuscule antenna in the tag need only be an extremely thin wire or printed coil, enough to pick up the transmitter's radio signal and, at the same time, induce a little voltage to power the chip in the tag. Normally, this is just enough power to transmit its own data back to the base station. Thus the base station knows which car passed and when (plus other things like name, address, and maybe even speed. Big brother is watching!).

In a few cases the tag has its own little battery, and is then called an *active* tag.

D. Nightingale and C. Spencer, *A Kitchen Course in Electricity and Magnetism,*   161
DOI 10.1007/978-3-319-05305-9, © Springer International Publishing Switzerland 2015

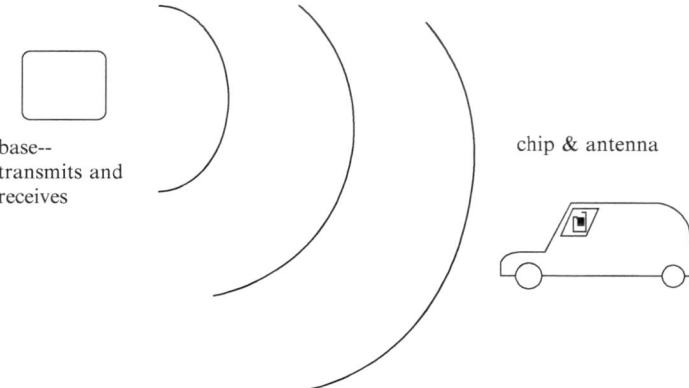

Fig. A.3 Cars approach the toll-taker's base which is transmitting a radio beam. This beam hits the car's windshield chip's tiny antenna, and induces a voltage—enough to power the chip and make it transmit a weak signal back to the base station. (This is not reflection of radio waves; it is retransmission.)

Interestingly, in 2009, researchers at Bristol University attached even more miniaturized tags onto the backs of ants and learned about their travels.

Note that keyless entry for a car is also by radio waves, of wavelength ~1 meter, although some older German cars used infrared wavelengths (which would be a fraction of a millimeter.) Either way, it's electromagnetic radiation, as in the spectrum of Fig. 3.46 on p. 116.

Also, general motion detectors for residences may use (short) radio waves, but other methods use heat radiation (*i.e.,* infrared) from warm bodies, such as automobile engines, people, and other animals.

Note on Other ID Methods (*e.g.,* Using Smart Phones)

Bar codes (Fig. A.4) which are ubiquitous for identifying grocery items, library books, and so on, where scanning across them with a light beam yields a sequence of reflections, can be translated into a series of small voltages, by means of our old friend the photodiode.

Fig. A.4 A typical Universal Product bar Code (UPC) on a box of ziti. Our photograph has been expanded horizontally for a clearer view of the differing bar widths

Basic bar codes were invented by **Bernard Silver** (1924–1963) and **Norman Woodland** (1921–2012) both then students at Drexel Institute of Technology in Philadelphia, in 1948, and were further developed by many others.

There are different types of code, and in the table below we have only given the bar thicknesses for the common UPC system (Universal Product Code). The bar has four varying widths of black stripes and white stripes, not dissimilar to Morse code dots and dashes extended downwards (bars). The light will be reflected from the white bars, and absorbed by the black bars, thus giving us voltage pulses from a photodiode (or photocell). The all-important relevant numbers, printed underneath, can also be keyed in should the scanner fail.

How to read a typical bar code? Neglecting for a moment the two parallel bars at the extreme ends of Fig. A.4, notice the (slightly taller) White–Black–White–Black. Their relative widths are fairly easily seen to be 3-2-1-1. From the table below, this corresponds to a 0 (which unfortunately is printed way over to the bottom left). If you don't like that, have a look at the first "7" on the left, and its thick black bar directly above it. Again referring to the table below, notice that the UPC code for "7" is 1-3-1-2, which means 1 white thickness, followed by 3 black thicknesses, 1 white thickness, and finally 2 thicknesses of black.

The center of the bar code is given by two thin black lines (here between the 8 and the 2), and the right hand "7" has the same pattern as the first "7" except that whites are now blacks. Of course, what all these numbers mean is a different matter—*e.g.*, the collection of numbers on the left of the two central lines is usually the manufacturer, such as the Coca-Cola company or Springer Publishing. As we emphasize, there are many different types of codes, and we only give here the key to the numerals used in the common UPC system:

0	3-2-1-1
1	2-2-2-1
2	2-1-2-2
3	1-4-1-1
4	1,1,3,2
5	1-2-3-1
6	1-1-1-4
7	1-3-1-2
8	1-2-1-3
9	3-1-1-

Two-dimensional bar codes (as in Fig. A.5) are somewhat different. They can be read by taking a photograph of the code with a so-called "smart-phone," which has a camera, and interprets the resulting small voltages coming this time from the CCD rather than a diode. Note that this is to do with visible light, therefore, not radio waves.

These *two-dimensional* codes (sometimes called "QR" or Quick Response codes) were created by the automotive industry in Japan in 1994. They are increasingly used worldwide for all kinds of things. An example might be "Take a flyer?"—no, photograph the [2D] bar code (Fig. A.5), use the free "app" to decode the CCD's image, and store the results (perhaps the flyer itself) in the phone's memory.

Fig. A.5 A typical 2-dimensional bar code, carrying more information than a 1-dimensional bar code. This pattern is photographed, and the digital information is interpreted by an "app" in the camera. They are used widely—for example by realtors, perhaps wishing to give details of a house for sale, and so on

Appendix D: E-Ink

E-Ink Display Screens

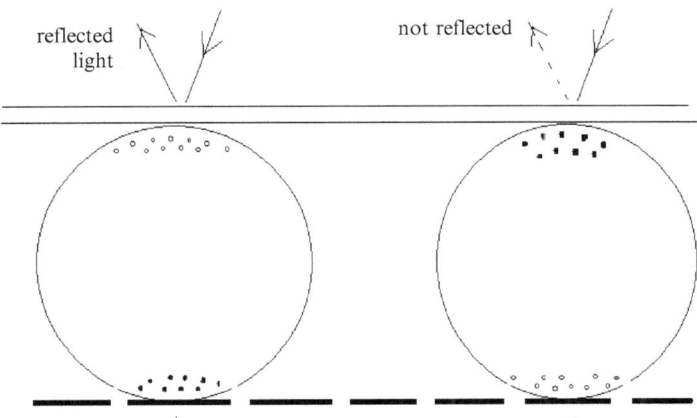

Fig. A.6 E-ink: If the *black particles* (negative) are at the *top* of a capsule, as in the RHS capsule, light will not be reflected; if *white particles* (positive) are at the *top* of a capsule, light will be reflected

Besides LCD screens (see p. 132) there is another important type, developed in the 1990s. It is the **E-ink** laminate, which may be attached to any surface (such as glass or even paper) and it is used now in many devices, especially e-book readers such as Kindle and others.

Developed by a professor and his students at MIT, the 1996 patent is held by **Dr. Joseph Jacobson** and **Barrett Comiskey**. The laminate is readable even in direct sunlight, and is as clear as the print on a page of paper. It uses minuscule white particles (charged positively) and black particles (charged negatively), and the basic idea is shown in Fig. A.6.

The minuscule particles are inside tiny capsules. When light is to be reflected from a capsule a (+) voltage from the bottom repels white particles, and so on.

Once again, we see that the governing idea is simply the Coulomb repulsion/ attraction discussed in Chap 1.

D. Nightingale and C. Spencer, *A Kitchen Course in Electricity and Magnetism*,
DOI 10.1007/978-3-319-05305-9, © Springer International Publishing Switzerland 2015

Appendix E: Touchscreens

Touchscreens have been in development since the 1960s. They are now in use in ATMs, i-phones, supermarkets, and so on, and are increasingly replacing keyboards.

We mention here two electrical methods that many touchscreens utilize—*Resistive,* and *Capacitive*—although there are also other very interesting methods which involve ultrasonics and infrared radiation. Below are sketches of the ideas behind Resistive and Capacitive methods respectively:

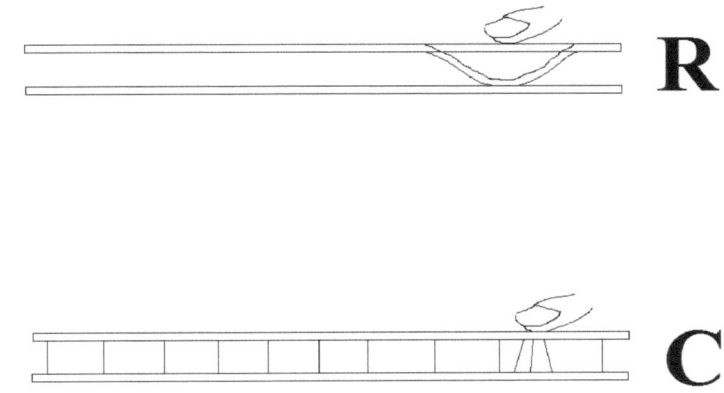

Fig. A.7 Two separate methods for touchscreens

In the "R" method a thin resistive material is sandwiched between two very thin glass plates, which have been made conducting by means of extremely thin coatings. (In Fig. A.7 the screen picture would be below either the "R" or the "C.") When the finger touches the top plate it pushes onto the other plate, and so obviously reduces, or even nullifies, the resistance at that spot. The location of that spot may be defined by tiny grid wires, x and y, (not shown) embedded in each of the plates.

In the "C" method the two conductive plates may be thought of as a Leyden jar (p. 16). The plates can this time be rigid, and note that there is of course an E-field between the plates, as there is with any capacitor, or between any separated charges.

D. Nightingale and C. Spencer, *A Kitchen Course in Electricity and Magnetism,*
DOI 10.1007/978-3-319-05305-9, © Springer International Publishing Switzerland 2015

Bringing a finger (an insulating rod or plastic pen is no good) to one part of a plate will slightly discharge that part (see the section on electrostatics, particularly p. 19), and so will change the local E-field. This will give rise to a minuscule voltage change at that location (again defined by x and y values, with grid wires not shown).

The reader may wonder why either of the top plates isn't at all the same potential. The answer lies in the design of the conductive plates, in tiny sections. Also, the whole thin two-plate assembly (for either the "R" or the "C" methods) has to be transparent enough for the computer screen picture to come through.

Appendix F: Formulae

Disclaimer! This Appendix is *only for those few* who have some knowledge of mathematics, and who *might* be browsing through the book in conjunction with an introductory calculus-based text. We have cross-referenced these standard physics equations with our "kitchen" descriptive pages, and the symbols used are typical throughout physics literature.

$\lvert\mathbf{F}\rvert = \text{const}\left(\dfrac{q_1 q_2}{r^2}\right)$	Coulomb's law	(p. xii)
$p = 2q\mathbf{l}$	Dipole definition, where [2l] is the separation between the charges	(p. 9)
$U = qV$	Definition of P.E. or Potential Energy, U	(p. 40)
$I = \dfrac{dq}{dt}$	Definition of current	(p. 29)
$\text{Power} = \text{V I} = \text{V}^2/\text{R} = \text{I}^2\text{R}$	Electrical power	(p. 55)
$V = IR$	Ohm's law	(p. 52)
$R = \varrho L/A$	Resistance, and resistivity (ϱ) [with table]	(pp. 50, 124)
$\sum_j I_j = 0$	Kirchhoff's first law	(p. 56)
$\sum_j V_j = 0$	Kirchhoff's second law	(p. 56)
$R = R_1 + R_2 + \cdots$	Resistors in series	
$\dfrac{1}{R} = \dfrac{1}{R_1} + \dfrac{1}{R_2} + \cdots$	Resistors in parallel	(p. 55)
$V = V_0 \sin(2\pi f t)$	Sine wave	(p. 73)
$P \propto V^2 \propto \langle\sin^2\rangle = 1/2$	Thus AC voltage needs "root 2"	(p. 74)
$\mu I = \oint \mathbf{B}.\,\mathbf{dl}$	Oersted's result	(p. 87)
Inductance of a solenoid	$L = \mu N^2 A/l$	(p. 89)
$\mathbf{F} = q[\mathbf{E} + (\mathbf{v}\times\mathbf{B})]$	Lorentz force	(p. 92)
or, for current only,	$= I(\mathbf{l}\times\mathbf{B})$	where l is the wire length
E.m.f. $= -N(d\varphi/dt)$	Faraday's law	(pp. 106, 108)
	where $N = $ # of conductors, and flux $\varphi = \mathbf{B}.\mathbf{A}$	(p. 109)
$c = f\lambda$	Speed of em wave	(p. 114)
		(continued)

D. Nightingale and C. Spencer, *A Kitchen Course in Electricity and Magnetism*,
DOI 10.1007/978-3-319-05305-9, © Springer International Publishing Switzerland 2015

Maxwell's equations:

$\nabla \cdot \mathbf{B} = 0$	No monopoles	(p. 79)
$\nabla \cdot \mathbf{D} = \varrho$	Gauss' law	(p. 18)
$\nabla \times \mathbf{E} = -\partial\mathbf{B}/\partial t$	Faraday's law	(pp. 106, 108)
$\nabla \times \mathbf{H} = \mathbf{J} + \partial\mathbf{D}/\partial t$	Ampere's law	(p. 88)
	where $\mathbf{D} = \varepsilon\mathbf{E}$, $\mathbf{B} = \mu\mathbf{H}$, $|\mathbf{J}| = I/A$	
$\mathbf{E} = -\nabla V$	Field is gradient of potential	(p. 143)

Glossary

Alkaline Nonacidic, with pH >7. Having a low concentration of H ions. Broccoli and cabbage are examples of alkaline foods.

Alkaline battery A battery with an alkaline electrolyte, such as potassium hydroxide. Alkaline batteries were developed in the 1960s and have a long shelf life. They are commonly used in AA, AAA, C, D, and 9 V types.

Ampere Unit of electrical current, meaning 1 Coulomb of charge passing any point in a conductor per second.

Astable Not stable; capable of oscillating between two states.

Atom The smallest particle of a chemical element, consisting of a positive nucleus surrounded by orbiting electrons.

Aurora borealis Literally, northern lights. Luminous streamers near either of the earth's poles.

Bar code A set of stripes of varying widths that can be read by a computer.

Bias A steady voltage (typically applied to an electronic device) that can be adjusted to change the way the device operates.

Breadboard A board that can receive the pins of components, for testing a circuit.

Capacitor A device consisting of two conducting plates separated by an insulator, used to store charge.

Charge Two distinct types exist in nature—positive and negative. Electrons, for example, have a negative charge, and protons have an equal and opposite charge to this.

Commutator Device for reversing the direction of an electric current in a motor.

Condenser Older word for *capacitor*.

Conductor A material that conducts electricity.

Coulomb Measure of electric charge.

Current Electric charge passing per unit time.

Depletion zone The variable region in a transistor where there are no charge carriers left, thus inhibiting current.

Diamagnetic A substance is diamagnetic if it is repelled by a magnetic field.

Diode A device that will allow current to pass in one direction only.

Dipole In electrostatics, a plus and a minus charge separated by a small distance; in magnetism, a magnet.

Eddy current Small circular electric currents induced in conductors.

D. Nightingale and C. Spencer, *A Kitchen Course in Electricity and Magnetism*,
DOI 10.1007/978-3-319-05305-9, © Springer International Publishing Switzerland 2015

Electrolysis Chemical separation, and deposition, by a direct electric current through an ionized substance, usually liquid or molten.

Electrolyte Electrolytes may be acids, bases, or salts, and have the property of passing electric current while being decomposed by the current.

Electrometer An instrument that measures voltage without drawing any current.

Electron Stable negatively charged unit of electricity, having a mass of about 1/1840 of a hydrogen atom.

Electrophorus A large conducting plate, such as a pie plate, with an insulating handle, used for transferring large charges, typically from a rubbed plastic sheet as a source. Invented by Johannes Wilcke in Sweden in the 1760s, and improved by Volta.

Electroscope Instrument for detecting electric charge.

Emf Electromotive force, measured in Volts.

Energy The ability to do work; in electricity, charge multiplied by Voltage.

Farad Measure of capacity, defined as 1 Coulomb/Volt. This is a very large unit, and the more common unit, the μF, or μFd, is one millionth of this.

Ferromagnetic Substances with large magnetic permeability, and thus easily magnetized. This property is lost at certain high temperatures, called the Curie temperature.

Field In electricity, the magnitude and direction of a force on a test charge; in magnetism, the magnitude and direction of a force on a test charge moving at unit velocity.

Flux The total amount of field lines passing through an area.

Galvanometer A very sensitive current meter.

Gedanken experiment A thought experiment.

Incandescent Glowing red hot

Inductance Currents "induce" magnetic fields around them; the effect is more pronounced with a coil. (See also *self-inductance*.)

Induction cooking The heating of a metal pot from eddy currents

Insulator A substance in which charges do not move when under the influence of an electric field.

Ion A charged atom (or molecule).

Joule heating Resistive heating in a wire, strongly proportional to the current and linearly proportional to the resistance.

kilowatt 1000 *Watts.*

LCD Liquid Crystal Display. The majority of flat screen TVs, as well as many clocks and watches, use LCDs. Such displays have a back light, and a liquid crystal between crossed polarizers which can rotate the plane of polarization because of small voltage changes, thus varying the amount of light that gets through.

Liquid crystal A liquid, in which the molecules are in a regular array, like that of a crystal.

Litz wire Stranded wire and thus flexible.

Magnetohydrodynamics Motion in a liquid due to a magnetic field.

Micro Prefix meaning one millionth.

Microwaves Electromagnetic waves of very short wavelength in the range (0.30–0.01) meters. (Microwave ovens operate at 0.12 m, which is about 5″.) Microwaves cannot pass around hills and mountains as longer wavelengths can.

Milli Prefix meaning one thousandth.

Molecule The smallest unit of a compound of atoms bonded together.

Neutron An electrically neutral particle about the same mass as a proton.

Ohm Unit of resistance; a Volt/Amp.

Paramagnetic Substance whose atoms have a permanent magnetic moment; substance having a permeability >1.

Pauli exclusion principle No two electrons in an atom may have the same four quantum numbers; of importance in the building up of the Periodic Table.

Permeability The ratio of the magnetic field inside a substance to the external field causing it. Usually very low, except in substances like iron.

pn junction Junction between p-type and n-type semiconductors.

Polarization Restriction of the plane of the electric (or magnetic) field of an electromagnetic wave to essentially one plane. In general, electromagnetic waves are unpolarized.

Potential In electrostatics, the value of the voltage at a point. It is directly related to the value of the electric field at that point.

Potential energy Charge times potential. A charge has the ability to do work whenever it is situated in a region where a potential exists, or equivalently, where an electric field exists.

Power Energy divided by time. (E.g., a watt is a joule/second. Also, one Horse Power is 746 Watts.)

Proton Positively charged particle 1836 times heavier than, and of equal and opposite charge to, the electron. Constituent of all atomic nuclei.

Radioactive Emitting alpha, and/or beta and/or gamma rays from the nucleus of an atom.

RFID Radio frequency identification.

Relay A magnetic switch triggered by a small current, usually to enable a larger current.

Resistance The property of impeding direct current flow. Specifically, the ratio of the potential difference across the ends of a conductor to the current flowing through the conductor. The practical unit of resistance is the *Ohm*.

Resonate Oscillate with the same frequency as the causative frequency.

Self-inductance A current through a conductor causes a magnetic field; if this current changes so will the magnetic field. The changing field will then induce a voltage back in the conductor.

Semiconductor A substance with resistivity considerably greater than that of a conductor and less than that of an insulator.

Short circuit A direct connection across the terminals of a battery or other voltage device.

Solar cell A device that produces voltage from sunlight.

Solenoid A coil of wire with length usually larger than radius. When a current is passed through a solenoid a fairly uniform magnetic field is formed inside, parallel to the coil's axis.

Spectrum In electromagnetism, all wavelengths (or frequencies) are possible.

Transducer A device that takes energy in one form and lets it off in another form. For example, a microphone, changing sound energy into electrical energy.

Transformer Two adjacent coils, wherein a changing voltage on one causes a changing (usually different) voltage on the other, with no change in frequency.

Transistor A device made from semiconductor material, used to control currents. The most common transistor has two closely spaced metallic electrodes, the Emitter and Collector, and the current between these is controlled by a third electrode, called the Base. There are other types of transistor, for example the Field Effect Transistor (FET), which similarly controls current, but via an electric field close to its third electrode, which is now called a Gate rather than a Base.

Volt Unit of electromotive force and/or potential difference.

Watt Unit of power, being 1 joule divided by 1 s.

Index

A

Ac. *See* Alternating current (AC)
Alfven, H., 103
Alternating current (AC), 67, 70, 72–76, 91,
    100, 109, 110, 116, 117, 166
Ampere, A.M., 29, 88
Amplifier, 17, 63, 103, 123
    transistor, 136–139
Amps, 29, 32, 38–40, 51–54, 56, 58, 61, 65, 66,
    68, 70, 75
Andersen, H.C., 87
Angle of dip, 82, 86
Arago, D.F.J., 119, 121
Arago's disk, 84, 119–121
Atom, 2, 5–10, 23, 25, 33, 42, 83, 88, 124–126
Aurora borealis, 86

B

Back emf, 91
Bar code readers, 64
Bar codes, 64, 161–164
Bardeen, J., 123
Battery, 12, 30–33, 35–43, 49–54, 56–67,
    70–72, 87, 88, 94, 98, 99, 102–104, 108,
    117, 121, 127–129, 134, 137, 140, 141,
    143, 144, 146, 148, 151–153, 161
Bellini, S., 86
B field. *See* Magnetic field
Blakemore, R., 86
Bohr, N., 7
Brushes, motor, 95
Busbar, 145

C

Camenzind, H.R., 150–151
Capacitors, 16–17, 19, 61, 63, 116, 142–153,
    157, 167
    charging and discharging, 144–149, 153

Carbon dating, 8
Cathode ray tubes (CRTs), 72, 101, 102
Cell, 12, 31–34, 36, 39, 56–64, 92, 108, 128, 157
Charge detector, 2, 11
Circuit-breaker, 42
Coil, 43, 44, 52, 67–73, 87–91, 96–101,
    108–109, 116, 120–122, 157, 161
Comiskey, B., 165
Commutator, 98–100
Compass, 9, 33, 43, 44, 54, 67, 77, 79–82,
    84–90, 119, 120
Conductor, 4, 6, 17, 23–25, 27, 97, 102, 107,
    115, 117, 121, 123, 150, 166
Core, iron/steel, 90, 109
Coulomb's law, 7, 18, 169
CRT. *See* Cathode ray tubes (CRTs)
Crystal set, 63, 116
Curie, J., 47
Curie, P., 47, 48
Current(s), 12, 16, 29–76, 82, 87–94, 98–105,
    107, 108, 113, 117–123, 127–130,
    133–135, 137–142, 146–149, 158, 169
    adjacent, 100–101
    meter, 32, 33, 42–45, 61, 67, 70, 107

D

Davy, H., 33–34, 41, 106
DC. *See* Direct current (DC)
de Maricourt, P., 77
de Coulomb, C.A., 14
Democritus, 6
Depletion region, 126
Diamagnetism, 83
Dielectric, 150
Diode, 49, 55, 56, 61–63, 67, 70, 75, 111,
    112, 116, 117, 123–130, 133, 135, 150,
    157, 164
Dipole, 9–10, 12, 78, 106, 169
Dirac, P., 79

D. Nightingale and C. Spencer, *A Kitchen Course in Electricity and Magnetism*,
DOI 10.1007/978-3-319-05305-9, © Springer International Publishing Switzerland 2015

Direct current (DC), 60, 67, 72–74, 91, 100,
    109, 110, 112, 116, 158
Dufay, C., xi

E
Earth's magnetic field, 80, 85, 86
Eddy currents, 84, 119–121
Edison, T.A., 41, 72, 73
E-ink display screens, 132, 165
Einstein, 6, 29, 57, 94, 102
Electric charges, 2, 8, 17, 29, 30, 39,
    86, 113
Electric current, 16, 29, 87, 121
Electric field (E field), 18–19, 78, 113, 116,
    141–143, 167, 168
Electric motor, 67, 93, 109, 121, 157
Electrode, 34–36, 38, 103, 104
Electrolysis, 33, 104
Electrolytes, 33, 36
Electrolytic capacitor, 144, 152
Electromagnetic induction, 72
Electromagnetic radiation, 105, 106, 162
Electromagnetic spectrum, 104, 106, 114–115
Electromagnetic waves, 113
Electro-motive force (EMF), 107
Electron, 4, 5, 7, 8, 10, 14, 16, 19, 23, 24, 31,
    38, 39, 41, 42, 50–54, 56, 57, 60, 63, 72,
    80, 83, 86, 88, 101–102, 105, 125–127,
    130, 133, 135
Electronic valves, 63
Electroplating, 34–35
Electroscope, 13–16, 19–24, 43, 62,
    140–143
    absolute, 14, 140–143
Electrostatics, 1–27, 56, 79, 132, 165
EMF. *See* Electro-motive force (EMF)
Energy, 16, 38–41, 56, 58, 73, 106, 116, 130,
    136, 143, 150, 151, 166
Euler, L., 148
Euler's number, 148
Exponential buildup, 145
Exponential decay, 145, 149

F
Farad, 148
Faraday cage, 27
Faraday, M., 17, 27, 32, 34, 83, 106, 109, 119
Faraday's law, 67, 106–109, 117, 119, 121,
    169, 170
Ferromagnetism, 83–84
Field effect transistor (FET), 140–142, 158

Flat screen TV, 132
Forward bias, 128
Franklin, B., 25, 26
Franklin's lightning bells, 24–27
Frequency, 73, 74, 105, 114, 116, 120, 130,
    153, 161
Fuse, 42, 66

G
Gain, 31, 38, 139
Galvani, L., 30
Gauss, J.C.F., 18
Gauss meter, 103
Gedanken, 79, 101
Generator, household, 121–122
Gilbert, W., 77, 85
Gray, S., 24

H
Hall, E., 102
Hall effect, 102–103
Heat, 8, 16, 41, 42, 44, 46, 53, 54, 60, 73, 80,
    106, 121, 141, 162
Helmholtz, H., xii
Henry, J., xi, 90, 91
Hertz, H., 57, 73, 161
House-alarm, 91–92

I
Inducing charges, 21–23
Inductance, 91, 118, 166
Induction cooking, 120
In parallel, 33, 55, 64–66, 68, 69, 166
In series, 32, 33, 36, 43, 55, 56, 59, 60, 64, 65,
    67, 68, 75, 76, 92, 94, 104, 129, 132,
    147, 157, 169
Insulator, 4, 6, 9, 20, 23–24, 123, 127, 135,
    150, 172
Integrated circuit (IC), 123–153
    555 timer chip, 149–152, 158
Inverse square law,
Ions, 8, 33, 60–61
Isotope, 8, 149

J
Jacobson, J., 165
Joule heating, 55, 73, 121
Joule, J.P., 16, 39
Joules, 16, 40

K
Kilowatt hours (kWh), 40
Kirchhoff, G.R., 56

L
Laser diode, 130
LCD. *See* Liquid crystal displays (LCDs)
LED. *See* Light-emitting diodes (LEDs)
Lenz, H., 118
Lenz's law, 118
Leyden jar, 16–21, 30, 167
Light-emitting diodes (LEDs),
      49–50
Lightning, 24–27, 33
Lines of force, 18, 84, 86
Lippmann, G., 48
Liquid crystal displays (LCDs), 101, 131,
      132, 165
Litz wire, 75
Lodestone, 77–79, 87
Lorentz force, 93–95, 101, 102,
      104–106, 169
Lorentz, H.A., 93
Loudspeaker, 23, 43, 48, 136, 137, 139

M
Magnet, 9, 43, 67, 68, 72, 77–87, 89,
      91–96, 100, 101, 103, 106–108,
      116–122
Magnetic bacteria, 86
Magnetic domains, 81
Magnetic field, 29, 78–80, 83–91, 93, 95,
      99–105, 107–109, 113, 116–118, 121
Magnetic flux, 109
Magnetic shielding, 84–85
Magnetic tape, 84, 86–87
Magnetohydrodynamics, 103–104
Magnetron, 105, 106
Maxwell, J.C., 73, 79, 132
Meter
      analog, 46, 48, 67, 144–145
      digital, 36, 46, 67, 144, 145
Meter, multi-, 32, 36, 37, 56, 66–68, 70, 72,
      108, 110, 144, 145, 148, 155
Metronome circuit, 152–153
Microphone, 48, 117, 136, 137, 139
Microwave ovens, 104–106
Microwaves, 105–106, 113
Molecules, 5, 6, 9–11, 24, 25, 86, 106

N
Neodymium magnets, 119
Neutron, 7, 8
Newton, I., 114
Nollet, l'Abbe, 16

O
Oersted, H.C., 87
Ohl, R.S., 57
Ohm, G.S., 50, 51
Ohms, 46, 50–52, 55, 56, 65, 66, 68–72, 75, 76,
      124, 128, 129, 134, 135, 138, 148, 155
Ohm's law, 51–56, 65–68, 70, 72, 75,
      129, 166

P
Paramagnetism, 83
Peltier effect, 47
Peltier, J.-C., 47
Permeability, 84, 85
Photodiode, 58–61, 64, 162, 163
Photoelectric effect, 57
Piezo-electric effect, 47
Pixel, 131
pn junction, 57, 58, 126–128, 130, 170
Polarization, 24, 25, 132
Polarized wave, 113
Potential, 39, 40, 129, 142–143, 146, 147,
      168, 170
      difference, 40, 102, 108, 146
      divider, 71, 128
      energy, 39, 143, 169
Potentiometer, 128, 157
Power, 24, 38–40, 47, 51, 53–55, 58, 63, 66,
      67, 73, 75, 121, 136, 161, 162, 169
Proton, 7, 8
pV cells, 58–61, 63, 64

Q
Quiescent level, 137, 139

R
Radioactive half life, 149
Radio frequency identification (RFID), 114,
      161–164
Radio, kitchen, 115–117, 161

Radio waves, 73, 104–106, 113–116, 161,
        162, 164
Rechargeable batteries, 60, 61
Reed switch, 91, 92
Relay, 90, 91, 151
Resistance, 50–58, 61, 65–73, 76, 89, 107, 123,
        129, 133, 134, 136–138, 145, 167, 169
    internal, 52, 67
Resistivity, 121, 123, 124, 166
Resistor color codes, 155
Reverse bias, 127
RFID. *See* Radio frequency identification
        (RFID)
Right-hand rule, 93, 100

S
Schottky, W.H., 61
Seebeck effect, 44–47
Seebeck thermocouple, 47
Seebeck, T.J., 44
Semiconductor, 4, 24, 57, 58, 61, 123,
        130, 150
Shell, electron, 7–9, 38, 60, 125
Silver, B., 163
Skin effect, 27, 74–75
Solar cell(s), 56–61, 64, 157
Solar panel, 59
Solenoid, 88–91, 100, 118, 166

Spencer, P., 105
Swan, J.W., 41
Swipe-cards, 86–87

T
Tesla, N., 72
Thales, xi
Thomson, J.J., 72
Time constant T, 148, 153
Touchscreens, 167–168
Transducer, 48, 137, 153, 158
        piezo-electric, 136, 152
Transformer, 73, 108–112
Transistor, 7, 123–153, 158

V
van Musschenbroek, P., 16
Volt, 16, 37–40, 54, 70, 129, 130, 174
Volta, C.A.G., 30–33
Voltage drop, 56, 61, 130
Voltaic cell, 31–32

W
Watt, J., 40
Woodland, N., 163
Work, mechanical, 16

Authors Biography

David Nightingale and **Chris Spencer** met as graduate students at Imperial College (London, UK), more than 50 years ago and have remained close friends ever since, meeting face to face about every 10 years on average. David was a Professor of Physics, State University of New York, College at New Paltz, for 31 years; and Chris was Senior Physics Teacher, Belmont Abbey School, Hereford, UK, for 26 years.

The authors' photo was taken amidst the ruins of the medieval Llanthony Priory in the Marches where Wales borders England.

D. Nightingale and C. Spencer, *A Kitchen Course in Electricity and Magnetism,*
DOI 10.1007/978-3-319-05305-9, © Springer International Publishing Switzerland 2015